Synthesis of
Quantum Circuits
vs.
Synthesis of
Classical Reversible Circuits

Synthesis Lectures on Digital Circuits and Systems

Editor

Mitchell A. Thornton, *Southern Methodist University*

The *Synthesis Lectures on Digital Circuits and Systems* series is comprised of 50- to 100-page books targeted for audience members with a wide-ranging background. The Lectures include topics that are of interest to students, professionals, and researchers in the area of design and analysis of digital circuits and systems. Each Lecture is self-contained and focuses on the background information required to understand the subject matter and practical case studies that illustrate applications. The format of a Lecture is structured such that each will be devoted to a specific topic in digital circuits and systems rather than a larger overview of several topics such as that found in a comprehensive handbook. The Lectures cover both well-established areas as well as newly developed or emerging material in digital circuits and systems design and analysis.

Synthesis of Quantum Circuits vs. Synthesis of Classical Reversible Circuits
Alexis De Vos, Stijn De Baerdemacker, and Yvan Van Rentergem
2017

Boolean Differential Calculus
Bernd Steinbach and Christian Posthoff
2017

Embedded Systems Design with Texas Instruments MSP432 32-bit Processor
Dung Dang, Daniel J. Pack, and Steven F. Barrett
2016

Fundamentals of Electronics: Book 4 Oscillators and Advanced Electronics Topics
Thomas F. Schubert and Ernest M. Kim
2016

Fundamentals of Electronics: Book 3 Active Filters and Amplifier Frequency
Thomas F. Schubert and Ernest M. Kim
2016

Bad to the Bone: Crafting Electronic Systems with BeagleBone and BeagleBone Black, Second Edition
Steven F. Barrett and Jason Kridner
2015

Fundamentals of Electronics: Book 2 Amplifiers: Analysis and Design
Thomas F. Schubert and Ernest M. Kim
2015

Fundamentals of Electronics: Book 1 Electronic Devices and Circuit Applications
Thomas F. Schubert and Ernest M. Kim
2015

Applications of Zero-Suppressed Decision Diagrams
Tsutomu Sasao and Jon T. Butler
2014

Modeling Digital Switching Circuits with Linear Algebra
Mitchell A. Thornton
2014

Arduino Microcontroller Processing for Everyone! Third Edition
Steven F. Barrett
2013

Boolean Differential Equations
Bernd Steinbach and Christian Posthoff
2013

Bad to the Bone: Crafting Electronic Systems with BeagleBone and BeagleBone Black
Steven F. Barrett and Jason Kridner
2013

Introduction to Noise-Resilient Computing
S.N. Yanushkevich, S. Kasai, G. Tangim, A.H. Tran, T. Mohamed, and V.P. Shmerko
2013

Atmel AVR Microcontroller Primer: Programming and Interfacing, Second Edition
Steven F. Barrett and Daniel J. Pack
2012

Representation of Multiple-Valued Logic Functions
Radomir S. Stankovic, Jaakko T. Astola, and Claudio Moraga
2012

Arduino Microcontroller: Processing for Everyone! Second Edition
Steven F. Barrett
2012

Advanced Circuit Simulation Using Multisim Workbench
David Báez-López, Félix E. Guerrero-Castro, and Ofelia Delfina Cervantes-Villagómez
2012

Circuit Analysis with Multisim
David Báez-López and Félix E. Guerrero-Castro
2011

Microcontroller Programming and Interfacing Texas Instruments MSP430, Part I
Steven F. Barrett and Daniel J. Pack
2011

Microcontroller Programming and Interfacing Texas Instruments MSP430, Part II
Steven F. Barrett and Daniel J. Pack
2011

Pragmatic Electrical Engineering: Systems and Instruments
William Eccles
2011

Pragmatic Electrical Engineering: Fundamentals
William Eccles
2011

Introduction to Embedded Systems: Using ANSI C and the Arduino Development Environment
David J. Russell
2010

Arduino Microcontroller: Processing for Everyone! Part II
Steven F. Barrett
2010

Arduino Microcontroller Processing for Everyone! Part I
Steven F. Barrett
2010

Digital System Verification: A Combined Formal Methods and Simulation Framework
Lun Li and Mitchell A. Thornton
2010

Progress in Applications of Boolean Functions
Tsutomu Sasao and Jon T. Butler
2009

Embedded Systems Design with the Atmel AVR Microcontroller: Part II
Steven F. Barrett
2009

Embedded Systems Design with the Atmel AVR Microcontroller: Part I
Steven F. Barrett
2009

Embedded Systems Interfacing for Engineers using the Freescale HCS08 Microcontroller II: Digital and Analog Hardware Interfacing
Douglas H. Summerville
2009

Designing Asynchronous Circuits using NULL Convention Logic (NCL)
Scott C. Smith and JiaDi
2009

Embedded Systems Interfacing for Engineers using the Freescale HCS08 Microcontroller I: Assembly Language Programming
Douglas H.Summerville
2009

Developing Embedded Software using DaVinci & OMAP Technology
B.I. (Raj) Pawate
2009

Mismatch and Noise in Modern IC Processes
Andrew Marshall
2009

Asynchronous Sequential Machine Design and Analysis: A Comprehensive Development of the Design and Analysis of Clock-Independent State Machines and Systems
Richard F. Tinder
2009

An Introduction to Logic Circuit Testing
Parag K. Lala
2008

Pragmatic Power
William J. Eccles
2008

Multiple Valued Logic: Concepts and Representations
D. Michael Miller and Mitchell A. Thornton
2007

Finite State Machine Datapath Design, Optimization, and Implementation
Justin Davis and Robert Reese
2007

Atmel AVR Microcontroller Primer: Programming and Interfacing
Steven F. Barrett and Daniel J. Pack
2007

Pragmatic Logic
William J. Eccles
2007

PSpice for Filters and Transmission Lines
Paul Tobin
2007

PSpice for Digital Signal Processing
Paul Tobin
2007

PSpice for Analog Communications Engineering
Paul Tobin
2007

PSpice for Digital Communications Engineering
Paul Tobin
2007

PSpice for Circuit Theory and Electronic Devices
Paul Tobin
2007

Pragmatic Circuits: DC and Time Domain
William J. Eccles
2006

Pragmatic Circuits: Frequency Domain
William J. Eccles
2006

Pragmatic Circuits: Signals and Filters
William J. Eccles
2006

High-Speed Digital System Design
Justin Davis
2006

Introduction to Logic Synthesis using Verilog HDL
Robert B.Reese and Mitchell A.Thornton
2006

Microcontrollers Fundamentals for Engineers and Scientists
Steven F. Barrett and Daniel J. Pack
2006

Synthesis of Quantum Circuits vs. Synthesis of Classical Reversible Circuits

Alexis De Vos, Stijn De Baerdemacker, and Yvan Van Rentergem

ISBN: 978-3-031-79894-8   paperback
ISBN: 978-3-031-79895-5   ebook
ISBN: 978-3-031-79896-2   hardcover

DOI 10.1007/978-3-031-79895-5

A Publication in the Springer series
SYNTHESIS LECTURES ON DIGITAL CIRCUITS AND SYSTEMS

Lecture #54
Series Editor: Mitchell A. Thornton, *Southern Methodist University*
Series ISSN
Print 1932-3166   Electronic 1932-3174

Synthesis of
Quantum Circuits
vs.
Synthesis of
Classical Reversible Circuits

Alexis De Vos, Stijn De Baerdemacker, and Yvan Van Rentergem
Universiteit Gent

SYNTHESIS LECTURES ON DIGITAL CIRCUITS AND SYSTEMS #54

ABSTRACT

At first sight, quantum computing is completely different from classical computing. Nevertheless, a link is provided by reversible computation.

Whereas an arbitrary quantum circuit, acting on w qubits, is described by an $n \times n$ unitary matrix with $n = 2^w$, a reversible classical circuit, acting on w bits, is described by a $2^w \times 2^w$ permutation matrix. The permutation matrices are studied in group theory of finite groups (in particular the symmetric group $\mathbf{S}_n$); the unitary matrices are discussed in group theory of continuous groups (a.k.a. Lie groups, in particular the unitary group $U(n)$).

Both the synthesis of a reversible logic circuit and the synthesis of a quantum logic circuit take advantage of the decomposition of a matrix: the former of a permutation matrix, the latter of a unitary matrix. In both cases the decomposition is into three matrices. In both cases the decomposition is not unique.

KEYWORDS

quantum computing, reversible computing, unitary matrix, permutation matrix, group theory, matrix decomposition, circuit synthesis

Contents

Acknowledgments

The authors thank the European COST Action IC 1405 "Reversible Computation" for its valuable support.

Alexis De Vos, Stijn De Baerdemacker, and Yvan Van Rentergem
May 2018

CHAPTER 1

Introduction

Often, in the literature, conventional computers and quantum computers are discussed as if they belong to two separate worlds, far from each other. Conventional computers act on classical bits, say "pure zeroes" and "pure ones," and by means of Boolean logic gates, such as AND gates and NOR gates. The operations performed by these gates are described by truth tables. Quantum computers act on qubits, say complex vectors, and by means of quantum gates, such as ROTATOR gates and T gates. The operations performed by these gates are described by unitary matrices.

Because the world of classical computation and the world of quantum computation are based on such different science models, it is difficult to see the relationship (be it analogies or differences) between these two computation paradigms. In the present chapter, we bridge the gap between the two sciences. For this purpose, a common language is necessary. The common tool we have chosen is the representation by square matrices and the construction of matrix groups.

1.1 CONVENTIONAL COMPUTING

A large majority of computers are digital computers, in other words, computers based on a set of two numbers: $\{0, 1\}$. We call the mathematics based on these two numbers Boolean algebra [1, 2]. A Boolean variable or bit can thus take only two different values: either 0 or 1. We call $f(A_1, A_2, \ldots, A_n)$ a Boolean function of n independent Boolean variables A_1, A_2, ..., A_{n-1}, and A_n. It takes either the value 0 or 1, depending on the values of its arguments A_1, A_2, ..., A_n. This dependency is fully described by means of a truth table, which tells us which value f has for each of the 2^n different values of the (Boolean) vector $(A_1, A_2, \ldots, A_n)$.

In the present chapter, we will survey some properties of Boolean functions, necessary to well understand binary reversible logic circuits. First we will have a close look at the Boolean functions $f(A)$ of a single variable, then at the Boolean functions $f(A_1, A_2)$ of two variables, before discussing the Boolean functions $f(A_1, A_2, \ldots, A_n)$ of an arbitrary number of Boolean variables. Besides unambiguously recording a Boolean function by writing down its truth table, one may also fully define a Boolean function by means of a (Boolean) formula. There are many ways to write down such a formula. We will discuss three standard ways: the minterm expansion, the Reed–Muller expansion, and the minimal ESOP expansion [3].

1.2   BOOLEAN FUNCTIONS OF ONE VARIABLE

There exist only four Boolean functions $f(A)$ of a single Boolean variable A. Table 1.1 shows the four corresponding truth tables. However, two of these functions are not really dependent of A. They are constants:

$$f(A) = 0 \qquad \text{(Table 1.1a)}$$
$$f(A) = 1 \qquad \text{(Table 1.1b)}.$$

We thus have only two true functions (or proper functions) of A :

$$f(A) = A \qquad \text{(Table 1.1c)}$$
$$f(A) = \overline{A} \qquad \text{(Table 1.1d)}.$$

The former is called the IDENTITY function, the latter is called the NOT function. Here, we have introduced the following short-hand notation for the NOT function:

$$\overline{X} = \text{NOT } X .$$

Whereas X is called a letter, both X and $\overline{X}$ are called a literal.

Table 1.1: Truth table of the four Boolean functions $f(A)$: (a) the constant function 0, (b) the constant function 1, (c) the IDENTITY function, and (d) the NOT function

A	f
0	0
1	0
(a)	

A	f
0	1
1	1
(b)	

A	f
0	0
1	1
(c)	

A	f
0	1
1	0
(d)	

1.3   BOOLEAN FUNCTIONS OF TWO VARIABLES

There exist $2^4 = 16$ different Boolean functions of two variables.[1] Table 1.2 shows them all. However, some of these functions $f(A, B)$ are not truly functions of A and B. Two functions are independent of both A and B. They are constants:

$$f_0(A, B) = 0$$
$$f_{15}(A, B) = 1 .$$

---

[1]Besides using the notation $A_1, A_2, ..., A_n$ for the variables, we will also use the letters $A, B, C, ...$, whenever this is more convenient.

Another four functions are in fact functions of a single variable: f_3 and f_{12} are independent of B, whereas both f_5 and f_{10} are independent of A:

$$f_3(A, B) = A$$
$$f_5(A, B) = B$$
$$f_{10}(A, B) = \overline{B}$$
$$f_{12}(A, B) = \overline{A}.$$

This leaves only $16 - 2 - 4 = 10$ functions, which are truly dependent on both A and B. We call them "true functions" of A and B.

Table 1.2: Truth table of all 16 Boolean functions $f_j(A, B)$

AB	f_0	f_1	f_2	f_3	f_4	f_5	f_6	f_7	f_8	f_9	f_{10}	f_{11}	f_{12}	f_{13}	f_{14}	f_{15}
0 0	0	0	0	0	0	0	0	0	1	1	1	1	1	1	1	1
0 1	0	0	0	0	1	1	1	1	0	0	0	0	1	1	1	1
1 0	0	0	1	1	0	0	1	1	0	0	1	1	0	0	1	1
1 1	0	1	0	1	0	1	0	1	0	1	0	1	0	1	0	1

Three out of the ten true functions of two variables (i.e., f_1, f_7, and f_6) are well known: the AND function, the OR function, and the XOR function (a.k.a. the EXCLUSIVE OR). Table 1.3 gives the corresponding truth tables. We will use the following short-hand notations for these basic Boolean functions:

$$X Y = X \text{ AND } Y$$
$$X + Y = X \text{ OR } Y$$
$$X \oplus Y = X \text{ XOR } Y.$$

Table 1.3: Truth table of three basic Boolean functions: (a) the AND function, (b) the OR function, and (c) the XOR function

AB	f		AB	f		AB	f
0 0	0		0 0	0		0 0	0
0 1	0		0 1	1		0 1	1
1 0	0		1 0	1		1 0	1
1 1	1		1 1	1		1 1	0
(a)			(b)			(c)	

The remaining $10 - 3 = 7$ functions are considered a combination of the NOT, AND, OR, and XOR functions. For example,

$$f_2(A, B) = A \text{ AND } (\text{NOT } B) = A\overline{B} \, .$$

We observe that all three functions AND, OR, and XOR are commutative:

$$A \text{ AND } B = B \text{ AND } A$$

and similar identities for the other two functions. This is not the case for any function $f(A, B)$. For example, the function f_2 is not commutative:

$$f_2(A, B) \neq f_2(B, A) \text{ but } f_2(A, B) = f_4(B, A) \, .$$

1.4 BOOLEAN FUNCTIONS OF n VARIABLES

There exist 2^{2^n} different Boolean functions of n variables. Each is represented by a truth table, consisting of $n + 1$ columns and 2^n rows. Table 1.4 gives an example for $n = 3$. The function is one out of the $2^8 = 256$ possibilities. Among them, only 218 are true functions of the three variables A, B, and C. Among the 256 functions, 70 are so-called balanced functions, in other words, functions which have an equal number of 1's and 0's in the output column. The function of Table 1.4 is both true and balanced.

A truth table can be summarized by a single Boolean formula. However, as mentioned in Section 1.1, there are multiple ways to write a given table as a Boolean expression [3]. We will now present three of them.

Table 1.4: Truth table of a function $f(A, B, C)$ of three variables

ABC	f
0 0 0	0
0 0 1	1
0 1 0	1
0 1 1	0
1 0 0	1
1 0 1	0
1 1 0	1
1 1 1	0

1.4.1 THE MINTERM EXPANSION

From the truth table, one can immediately deduce the Boolean formula called the minterm expansion. For example, Table 1.4 immediately yields:

$$f(A, B, C) = \overline{A}\,\overline{B}\,C + \overline{A}\,B\,\overline{C} + A\,\overline{B}\,\overline{C} + AB\,\overline{C} \ .$$

It consists of the OR of different terms. There are between 0 and 2^n different terms present. Each term is called a minterm and consists of the AND of exactly n literals.

The algorithm for translating the truth table into the minterm expansion is straightforward: each row of the truth table with a 1 in the rightmost column (i.e., the output column) yields one minterm; the latter consists of an AND of all input letters, overlined if in the corresponding column appears a 0, not overlined if in the corresponding column appears a 1.

In the literature such an expansion is sometimes referred to as a "sum of products" because an OR in some way "resembles a sum" and an AND in a way "resembles a product." Also the abbreviation SOP is often used.

1.4.2 THE REED–MULLER EXPANSION

A fundamentally different expansion is the Reed–Muller expansion. It is obtained as follows: we apply to the minterm expansion the two identities

$$\overline{X} = 1 \oplus X$$
$$X + Y = X \oplus Y \oplus XY \ . \tag{1.1}$$

This leads to a XOR of ANDs. The result is subsequently simplified by applying the identities

$$X \oplus X = 0$$
$$0 \oplus X = X \ .$$

In case of our example (Table 1.4), we obtain

$$f(A, B, C) = A \oplus B \oplus C \oplus AB \ . \tag{1.2}$$

A Reed–Muller expansion is an example of an ESOP expansion, in other words, an "EXCLUSIVE-OR sum of products." Thus, just like the OR, the XOR function also is considered "a kind of sum."

In many respects, a Reed–Muller expansion of a Boolean function resembles the well-known Taylor expansion of ordinary calculus. Let us assume a function f of the real numbers x, y, and z. Then, the Taylor expansion around the point $(x, y, z) = (0, 0, 0)$ looks like

$$\begin{aligned}
f(x, y, z) = {}& c_{000} + c_{100}x + c_{010}y + c_{001}z + \\
& c_{110}xy + c_{101}xz + c_{011}yz + c_{200}x^2 + c_{020}y^2 + c_{002}z^2 + \\
& c_{111}xyz + c_{210}x^2 y + \cdots \ .
\end{aligned}$$

The Reed–Muller expansion of a function f of the Boolean numbers A, B, and C looks like

$$f(A, B, C) = c_{000} \oplus c_{100} A \oplus c_{010} B \oplus c_{001} C \oplus$$
$$c_{110} AB \oplus c_{101} AC \oplus c_{011} BC \oplus c_{111} ABC .$$

There are three main differences.

- The Reed–Muller coefficients c_{ijk} can have only two possible values: either 0 or 1.

- The exponents i, j, and k in the monomial (a.k.a. term or piterm) $A^i B^j C^k$ each can have only two different values: either 0 or 1. As a result:

- There are only a finite number of Reed–Muller terms.

We denote by $R(n)$ the maximum number of terms in the Reed–Muller expansion of an arbitrary Boolean function of n variables:

$$R(n) = 2^n . \tag{1.3}$$

1.4.3   THE MINIMAL ESOP EXPANSION

In the Reed–Muller expansion, NOT functions are not allowed.[2] If we do allow NOT operations, the "XOR of ANDs" expansion can be made shorter. The shortest expansion (i.e., the one with the minimum number of terms) is called the minimal ESOP expansion.

The minimal ESOP expansion is quite different from the two above expansions (i.e., the minterm expansion and the Reed–Muller expansion) in two respects:

- it is not unique: there may exist two or even more minimal ESOP expansions of a same Boolean function, and

- there is no straightforward algorithm to find the minimal ESOP expansion(s) (except, of course, exhaustive search).

The last fact explains why minimal ESOPs are only known for the Boolean functions with $n = 6$ or less [4, 5].

Our example function (Table 1.4) has two different minimal ESOPs:

$$f(A, B, C) = A \oplus C \oplus \overline{A} B$$
$$= B \oplus C \oplus A \overline{B} .$$

Whereas the Reed–Muller expansion (1.2) needs four terms, these minimal ESOPs contain only three terms; whereas the Reed–Muller expansion (1.2) needs five literals, these minimal ESOPs contain only four literals.

---

[2]In the present book we limit ourselves to so-called "positive-polarity Reed–Muller expansions." We thus ignore here "negative-polarity Reed–Muller expansions" and "mixed-polarity Reed–Muller expansions" [3].

We denote by $E(n)$ the maximum number of terms in the minimal ESOP expansion of an arbitrary Boolean function of n variables. In contrast to the simple $R(n)$ formula (1.3), we have only very partial knowledge [4–6] about this function $E(n)$:

$$
\begin{aligned}
E(1) &= 1 \\
E(2) &= 2 \\
E(3) &= 3 \\
E(4) &= 6 \\
E(5) &= 9 \\
E(6) &= 15 \\
E(n) &\leq 29 \times 2^{n-7} \text{ for } n \geq 7 \,.
\end{aligned}
\tag{1.4}
$$

1.5   GROUP THEORY

A very important mathematical tool for investigating Boolean functions is the group. For mathematicians, a group **G** consists of the combination of two things:

- a set $S = \{a, b, c, \ldots\}$, and

- an operation Ω (involving two elements of the set).

However, the set and the operation have to fulfill four conditions. Applying the infix notation for the bivariate function Ω, these conditions are:

- S has to be closed:
  $a \, \Omega \, b \in S$;

- Ω has to be associative:
  $(a \, \Omega \, b) \, \Omega \, c = a \, \Omega \, (b \, \Omega \, c)$;

- S has to have an identity element i:
  $a \, \Omega \, i = a$;

- each element of S has to have an inverse in S:
  $a \, \Omega \, a^{-1} = i$.

The number of elements in the set is called the order of the group.

We start with an example: the set of all 2^{2^n} Boolean functions of n variables (Section 1.4) forms a group with respect to the operation XOR (thus with Ω = XOR). Indeed all four conditions are fulfilled:

- If f_1 and f_2 are Boolean functions, then also $f_1 \oplus f_2$ is.

- If f_1, f_2, and f_3 are Boolean functions, then $(f_1 \oplus f_2) \oplus f_3$ equals $f_1 \oplus (f_2 \oplus f_3)$. Therefore we simply write $f_1 \oplus f_2 \oplus f_3$.

- There is an identity element i: the zero function 0. Indeed, if f is an arbitrary function, then $f \oplus 0 = 0 \oplus f = f$.

- If f is an arbitrary function, then f is its own inverse, as $f \oplus f = 0$.

In the particular case of $n = 1$, there are four functions f_j of one variable A (see Section 1.2): $f_0(A) = 0$, $f_1(A) = A$, $f_2(A) = \overline{A}$, and $f_3(A) = 1$. The 16 possible outcomes $f_j \oplus f_k$ are given by the so-called Cayley table: Table 1.5.

Table 1.5: Cayley table of the four Boolean functions $f_j(A)$

	f_0	f_1	f_2	f_3
f_0	f_0	f_1	f_2	f_3
f_1	f_1	f_0	f_3	f_2
f_2	f_2	f_3	f_0	f_1
f_3	f_3	f_2	f_1	f_0

A group counterexample is formed by the same set, however, combined with the OR operation. We have the following.

- If f_1 and f_2 are Boolean functions, then also $f_1 + f_2$ is so.

- If f_1, f_2, and f_3 are Boolean functions, then $(f_1 + f_2) + f_3$ equals $f_1 + (f_2 + f_3)$.

- There is an identity element $i = 0$. Indeed, if f is an arbitrary function, then $f + 0 = 0 + f = f$.

But, if f is an arbitrary function different from the zero function (thus with at least one 1 in the output column of its truth table), then there exists no function g such that $f + g = 0$. Thus the last group condition (existence of an inverse) is not fulfilled.

We give a second example of a group. The set S consists of merely two elements: the zero function 0 and one particular Boolean function f of n variables, with only one 1 in the truth table's output column. In other words, the minterm expansion of f consists of a single minterm. Therefore, we call it a minterm function. For a group, besides the set $S = \{0, f\}$, we also need a bivariate operation Ω. This operation again is the XOR function. The reader is invited to verify that all group conditions are fulfilled, for example, the first group condition, i.e., that $0 \oplus 0$, $0 \oplus f$, $f \oplus 0$, and $f \oplus f$ are in $\{0, f\}$. We may call the group a "minterm group." The order of the group is 2. Mathematicians call this group the symmetric group of degree 2, with notation $\mathbf{S}_2$. In general, the symmetric group of degree n consists of all possible permutations of n objects. It is denoted $\mathbf{S}_n$ and has order $n!$, that is, the factorial of n.

As an exercise, the reader is invited to consider the six balanced functions $f_j(A, B)$ in Table 1.2 and check whether the XOR operation leads to a group.

Below, we will drop the explicit rendering of the symbol Ω, by writing ab instead of $a \, \Omega \, b$. We also will call ab the product of a and b, even in cases where Ω is not a multiplication. Note that often ab is not the same as ba. Groups where, for all couples $\{a, b\}$, we do have $ab = ba$, are called commutative or Abelian groups. Most groups we will encounter in the present book are not Abelian. The group $\mathbf{S}_2$ is Abelian; the groups $\mathbf{S}_n$ with $n > 2$ are not Abelian.

Besides the group of all permutations on n objects, that is, $\mathbf{S}_n$ of order $n!$, we will encounter in the present book, is also the group of cyclic permutations of n objects. It is called the cyclic group, is denoted $\mathbf{C}_n$, has order n, and is Abelian.

Above, the set S consists either of functions or of permutations. A third example is a set S of matrices. Then, the operation Ω is the matrix multiplication.

A restricted set $\{g_1, g_2, \ldots, g_k\}$ of elements of a group $\mathbf{G}$ is said to generate $\mathbf{G}$ if all members of $\mathbf{G}$ can be written as a finite product of the elements of the set. The elements g_1, g_2, ..., and g_k are called the generators of $\mathbf{G}$. Such a set of generators usually is not unique. Surprising is the fact that many (even large) groups need only very few generators. The symmetric group $\mathbf{S}_n$ needs only two (well chosen) generators; the cyclic group $\mathbf{C}_n$ needs only one generator. For example, the group $\mathbf{S}_4$ is generated by the two matrices

$$g_1 = \begin{pmatrix} 0 & 1 & 0 & 0 \\ 1 & 0 & 0 & 0 \\ 0 & 0 & 1 & 0 \\ 0 & 0 & 0 & 1 \end{pmatrix} \quad \text{and} \quad g_2 = \begin{pmatrix} 0 & 1 & 0 & 0 \\ 0 & 0 & 1 & 0 \\ 0 & 0 & 0 & 1 \\ 1 & 0 & 0 & 0 \end{pmatrix}. \tag{1.5}$$

This means that all $4! = 24$ matrices of $\mathbf{S}_4$ can be written as products like g_1, g_2, g_1^2, $g_1 g_2$, $g_2 g_1$, g_2^2, g_1^3, $g_1^2 g_2$. The group $\mathbf{C}_4$ is generated by the single matrix

$$g = \begin{pmatrix} 0 & 1 & 0 & 0 \\ 0 & 0 & 1 & 0 \\ 0 & 0 & 0 & 1 \\ 1 & 0 & 0 & 0 \end{pmatrix}.$$

This means that all four matrices of $\mathbf{C}_4$ can be written as g, g^2, g^3, and g^4.

If $\{g_1, g_2, \ldots\}$ are generators of a group $\mathbf{G}$ and $\{k_1, k_2, \ldots\}$ are generators of a group $\mathbf{K}$, then the group generated by the union $\{g_1, g_2, \ldots\} \cup \{k_1, k_2, \ldots\}$ is called the closure of $\mathbf{G}$ and $\mathbf{K}$.

Two groups $\mathbf{G}$ and $\mathbf{H}$ are called isomorphic if they have the same order and one can find a 1-to-1 relationship $g \leftrightarrow h$ between the elements g of $\mathbf{G}$ and the elements h of $\mathbf{H}$, such that $g_1 \leftrightarrow h_1$ and $g_2 \leftrightarrow h_2$ automatically implies $g_1 g_2 \leftrightarrow h_1 h_2$. In other words, the two groups follow a same Cayley table (a.k.a. product table). We write $\mathbf{G} \cong \mathbf{H}$. The reader may easily verify the example $\mathbf{C}_2 \cong \mathbf{S}_2$. Another example of two groups isomorphic to one another will be given in Section 1.9.

For computations and experiments with (finite) groups, the computer algebra package GAP is recommended [7].

1.6   REVERSIBLE COMPUTING

The first step in bridging the gap between classical and quantum computation is replacing conventional classical computing by reversible classical computing. Whereas conventional logic gates are represented by truth tables with an arbitrary number w_i of input columns and an arbitrary number w_o of output columns, reversible logic gates are described by truth tables with an equal number w of input and output columns. We call w_i, w_o, and w the input width, the output width, and the width of the circuit, respectively.

Besides an equal input and output width, a reversible truth table has the property that all output rows are different, such that the 2^w output words are merely a permutation of the 2^w input words [8–10]. Table 1.6 gives an example of a conventional gate (i.e., the AND gate, with two input bits A and B and one output bit R), as well as an example of a reversible gate (i.e., a TOFFOLI gate, a.k.a. a controlled NOT gate, with three input bits A, B, and C and three output bits P, Q, and R). We call Table 1.6a (with $w_i = 2$ inputs and $w_o = 1$ outputs) an irreversible truth table because knowledge of the output value is not sufficient to recover the input values. Indeed, if we know that $R = 0$, this is not sufficient to trace what A and B "have been." We call Table 1.6b (with widths $w_i = w_o = 3$) reversible, because knowledge of the output values P, Q, and R is sufficient to recover the input values A, B, and C.

The reader may verify that the irreversible AND function is embedded in the reversible TOFFOLI function, as presetting in Table 1.6b the input C to logic 0 leads to the output R being equal to A AND B, as is highlighted by boldface. In the general case, any irreversible truth table can be embedded in a reversible truth table with $w = w_i + w_o$ or less bits [11].

Because the eight output words PQR of the reversible truth table are a permutation of the eight input words, automatically all three functions $P(A, B, C)$, $Q(A, B, C)$, and $R(A, B, C)$ are balanced. Their minimal ESOP expressions are

$$
\begin{aligned}
P &= A \\
Q &= B \\
R &= C \oplus AB \, .
\end{aligned}
$$

Because the 2^w output words of a reversible truth table are a permutation of the 2^w input words, we have to study in detail the group of the permutations of n objects. We denote this group by $\mathrm{P}(n)$ and recall that it is isomorphic to the symmetric group $\mathbf{S}_n$ of order $n!$.

The next step in the journey from the conventional to the quantum world is replacing the reversible truth table by a permutation matrix. As all eight output words $0\,0\,0, 0\,0\,1, \ldots$, and $1\,1\,0$ are merely a permutation of the eight input words $0\,0\,0, 0\,0\,1, \ldots$, and $1\,1\,1$, Table 1.6b can be

Table 1.6: Truth table of two basic Boolean functions: (a) the AND function, (b) the TOFFOLI function

ABC	PQR
0 0 0	**0 0 0**
0 0 1	0 0 1
0 1 0	**0 1 0**
0 1 1	0 1 1
1 0 0	**1 0 0**
1 0 1	1 0 1
1 1 0	**1 1 1**
1 1 1	1 1 0

(b)

AB	R
0 0	0
0 1	0
1 0	0
1 1	1

(a)

replaced by an 8×8 permutation matrix, i.e.,

$$
\begin{pmatrix}
1 & 0 & 0 & 0 & 0 & 0 & 0 & 0 \\
0 & 1 & 0 & 0 & 0 & 0 & 0 & 0 \\
0 & 0 & 1 & 0 & 0 & 0 & 0 & 0 \\
0 & 0 & 0 & 1 & 0 & 0 & 0 & 0 \\
0 & 0 & 0 & 0 & 1 & 0 & 0 & 0 \\
0 & 0 & 0 & 0 & 0 & 1 & 0 & 0 \\
0 & 0 & 0 & 0 & 0 & 0 & 0 & 1 \\
0 & 0 & 0 & 0 & 0 & 0 & 1 & 0
\end{pmatrix} .
\tag{1.6}
$$

An arbitrary classical reversible circuit, acting on w bits, is represented by a permutation matrix of size $2^w \times 2^w$. In contrast, a quantum circuit, acting on w qubits, is represented by a unitary matrix U of size $2^w \times 2^w$. Both kinds of matrices are depicted by symbols with w input lines and w output lines:

$$
\begin{array}{ccc}
A & \quad & P \\
B & U & Q \\
C & \quad & R .
\end{array}
$$

Invertible square matrices, together with the operation of ordinary matrix multiplication, form a group. The finite matrix group $P(2^w)$ consisting solely of permutation matrices is a subgroup of the continuous group $U(2^w)$ of unitary matrices. In Chapter 3, we will show a natural means to enlarge the subgroup to its supergroup, in other words, how to upgrade a classical computer to a quantum computer.

1.7   PERMUTATION GROUPS

Permutation groups take a special place in group theory because any finite group is isomorphic to some permutation group. A permutation group consists of a set of permutations together with the operation of cascading. Tables 1.7a and 1.7b show two different permutations of the eight objects 1, 2, 3, 4, 5, 6, 7, and 8. Table 1.7c gives the permutation resulting from the cascading of the previous two permutations. In order to deduce Table 1.7c from Tables 1.7a and 1.7b, we proceed as follows: the first row of Table 1.7a tells "1 is mapped to 2," whereas the second row of Table 1.7b tells "2 is mapped to 3." Together this yields "1 is mapped to 3," to be filled in, in the first row of Table 1.7c. We may equally well say: according to Table 1.7a, "2 is the image of 1," whereas according to Table 1.7b, "3 is the image of 2." Together this yields: "3 is the image of the image of 1," to be filled in, in the first row of Table 1.7c. Subsequently proceeding like this for all eight rows of Table 1.7a yields the full Table 1.7c.

Permutations of n objects can be represented by permutation matrices, i.e., $n \times n$ matrices with all entries either equal to 0 or equal to 1 and all line sums equal to 1. By definition, a line sum is either a row sum or a column sum. In a permutation matrix almost all entries are 0. In each row, one and only one entry equals 1; in each column, one and only one entry equals 1. For example, Table 1.7a is represented by the matrix

$$\begin{pmatrix} 0 & 0 & 1 & 0 & 0 & 0 & 0 & 0 \\ 1 & 0 & 0 & 0 & 0 & 0 & 0 & 0 \\ 0 & 1 & 0 & 0 & 0 & 0 & 0 & 0 \\ 0 & 0 & 0 & 1 & 0 & 0 & 0 & 0 \\ 0 & 0 & 0 & 0 & 0 & 1 & 0 & 0 \\ 0 & 0 & 0 & 0 & 1 & 0 & 0 & 0 \\ 0 & 0 & 0 & 0 & 0 & 0 & 1 & 0 \\ 0 & 0 & 0 & 0 & 0 & 0 & 0 & 1 \end{pmatrix} . \tag{1.7}$$

Each row of the correspondence Table 1.7a is translated into a corresponding column of the permutation matrix. For example, because 2 is mapped to 3, the second column has a 1 in the third row. Table 1.7c, being the cascade of the Tables 1.7a and 1.7b, can be written as a matrix product:

$$\begin{pmatrix} 1 & 0 & 0 & 0 & 0 & 0 & 0 & 0 \\ 0 & 0 & 1 & 0 & 0 & 0 & 0 & 0 \\ 0 & 1 & 0 & 0 & 0 & 0 & 0 & 0 \\ 0 & 0 & 0 & 0 & 0 & 0 & 0 & 1 \\ 0 & 0 & 0 & 0 & 0 & 1 & 0 & 0 \\ 0 & 0 & 0 & 0 & 1 & 0 & 0 & 0 \\ 0 & 0 & 0 & 0 & 0 & 0 & 1 & 0 \\ 0 & 0 & 0 & 1 & 0 & 0 & 0 & 0 \end{pmatrix} \begin{pmatrix} 0 & 0 & 1 & 0 & 0 & 0 & 0 & 0 \\ 1 & 0 & 0 & 0 & 0 & 0 & 0 & 0 \\ 0 & 1 & 0 & 0 & 0 & 0 & 0 & 0 \\ 0 & 0 & 0 & 1 & 0 & 0 & 0 & 0 \\ 0 & 0 & 0 & 0 & 0 & 1 & 0 & 0 \\ 0 & 0 & 0 & 0 & 1 & 0 & 0 & 0 \\ 0 & 0 & 0 & 0 & 0 & 0 & 1 & 0 \\ 0 & 0 & 0 & 0 & 0 & 0 & 0 & 1 \end{pmatrix} =$$

$$\begin{pmatrix} 0 & 0 & 1 & 0 & 0 & 0 & 0 & 0 \\ 0 & 1 & 0 & 0 & 0 & 0 & 0 & 0 \\ 1 & 0 & 0 & 0 & 0 & 0 & 0 & 0 \\ 0 & 0 & 0 & 0 & 0 & 0 & 0 & 1 \\ 0 & 0 & 0 & 0 & 1 & 0 & 0 & 0 \\ 0 & 0 & 0 & 0 & 0 & 1 & 0 & 0 \\ 0 & 0 & 0 & 0 & 0 & 0 & 1 & 0 \\ 0 & 0 & 0 & 1 & 0 & 0 & 0 & 0 \end{pmatrix}.$$

Observe that, in the above matrix product, the matrix (1.7) representing the permutation applied first is written to the right of the matrix representing the permutation applied afterwards. Indeed, mathematicians apply matrices from right to left. It is no surprise this sometimes leads to confusion and/or errors.

Table 1.7: Correspondence table of three different permutations of 8 objects

A	P	A	P	A	P
1	2	1	1	1	3
2	3	2	3	2	2
3	1	3	2	3	1
4	4	4	8	4	8
5	6	5	6	5	5
6	5	6	5	6	6
7	7	7	7	7	7
8	8	8	4	8	4
(a)		(b)		(c)	

As an example of a permutation group, we consider the group of all permutations of three objects 1, 2, and 3. It has order $3! = 6$. Its six elements are represented by the six permutation matrices

$$\begin{pmatrix} 1 & 0 & 0 \\ 0 & 1 & 0 \\ 0 & 0 & 1 \end{pmatrix}, \begin{pmatrix} 0 & 1 & 0 \\ 1 & 0 & 0 \\ 0 & 0 & 1 \end{pmatrix}, \begin{pmatrix} 0 & 0 & 1 \\ 0 & 1 & 0 \\ 1 & 0 & 0 \end{pmatrix},$$

$$\begin{pmatrix} 1 & 0 & 0 \\ 0 & 0 & 1 \\ 0 & 1 & 0 \end{pmatrix}, \begin{pmatrix} 0 & 1 & 0 \\ 0 & 0 & 1 \\ 1 & 0 & 0 \end{pmatrix}, \text{ and } \begin{pmatrix} 0 & 0 & 1 \\ 1 & 0 & 0 \\ 0 & 1 & 0 \end{pmatrix}. \tag{1.8}$$

They form the symmetric group $\mathbf{S}_3$. Above, the element $\begin{pmatrix} 1 & 0 & 0 \\ 0 & 1 & 0 \\ 0 & 0 & 1 \end{pmatrix}$ is the trivial permutation which maps each object to itself. In other words, by this permutation, no object is "moved." This element is the identity element i of the group. Its matrix representation is a diagonal matrix, with exclusively 1's in the diagonal: the 3×3 unit matrix.

1.8   A PERMUTATION DECOMPOSITION

If n is not prime, a powerful permutation decomposition exists [9, 12]. Let n be an integer, in other words, the number of objects in the set $\{1, 2, ..., n\}$. Let p be a divisor of n, that is, we have $n = pq$, with both p and q integers (greater than 1). We arrange the n objects into q rows, each of p objects, the arrangement being called a Young tableau [13]. We consider an arbitrary permutation a of the n objects. Figure 1.1a shows the permutation a as a mapping in the Young tableau for $n = 35$, $p = 7$, and $q = 5$. The tableau consists of the five sets $\{1, 2, ..., 7\}$, $\{8, 9, ..., 14\}$, ..., and $\{29, 30, ..., 35\}$, each of seven objects. We see here how 1 is mapped to 1, how 2 is mapped to 2, how 3 is mapped to 23, and so on.

Theorem 1.1   *Each permutation a can be decomposed as*

$$a = h_1 * v * h_2, \tag{1.9}$$

where both h_1 and h_2 only permute objects within rows of the Young tableau and where v only permutes objects within columns of the tableau.

Figures 1.1b, 1.1c, and 1.1d show the permutations h_1, v, and h_2 to be performed successively.

- The vertical permutation v (Figure 1.1c) is found as follows: the cycles of a are projected onto columns, yielding one or more vertical cycles.

- The horizontal permutation h_1 (Figure 1.1b) merely consists of horizontal arrows which map the arrow tails of vertical and oblique arrows in Figure 1.1a to the corresponding arrow tails of Figure 1.1c. Subsequently, additional horizontal arrows are added in order to form closed horizontal cycles.

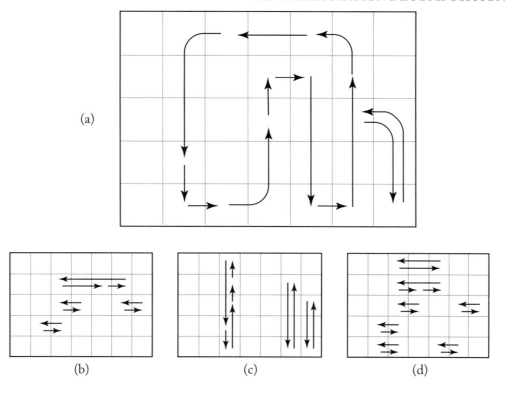

Figure 1.1: Decomposition of (a) an arbitrary permutation of 35 objects into (b) a first "horizontal" permutation, (c) a "vertical" permutation, and (d) a second "horizontal" permutation.

- Finally, the horizontal permutation h_2 (Figure 1.1d) simply equals $v^{-1} * h_1^{-1} * a$.

The fact that it is always possible to construct an appropriate vertical permutation v is a consequence of Birkhoff's theorem [14, 15] on doubly stochastic matrices, more specifically the integer case [16, 17] of Birkhoff's theorem.

We conclude: If an integer n can be factorized as $p \times q$, then any permutation of n objects can be performed by applying, subsequently,

- q permutations, each of p objects,

- p permutations, each of q objects, and

- q permutations, each of p objects.

A slightly more powerful property can be proved as follows.

Theorem 1.2   *In the horizontal permutation h_1, it is always possible to not permute objects in the upper row. Hence, if an integer n can be factorized as $p \times q$, then any permutation of n objects can be performed by applying, subsequently,*

- *$q - 1$ permutations, each of p objects,*

- *p permutations, each of q objects, and*

- *q permutations, each of p objects.*

   It is clear that Theorem 1.1 is a direct consequence of Theorem 1.2. Unfortunately, the stronger Theorem 1.2 lacks the beautiful symmetry of Theorem 1.1. Symmetry is somewhat restored by the existence of the following.

Theorem 1.3   *If an integer n can be factorized as $p \times q$, then any permutation of n objects can be performed by applying, subsequently,*

- *q permutations, each of p objects,*

- *p permutations, each of q objects, and*

- *$q - 1$ permutations, each of p objects.*

1.9   MATRIX GROUPS

The study of matrix groups is interesting because any finite group is isomorphic to some matrix group and many infinite groups as well. A matrix group consists of a set of square matrices together with the operation of matrix multiplication. Because matrix multiplication is not commutative, most matrix groups are not Abelian.

   The reader is invited to check that the following six 2×2 matrices form a group:

$$\begin{pmatrix} 1 & 0 \\ 0 & 1 \end{pmatrix}, \begin{pmatrix} 1/2 & -\sqrt{3}/2 \\ -\sqrt{3}/2 & -1/2 \end{pmatrix}, \begin{pmatrix} 1/2 & \sqrt{3}/2 \\ \sqrt{3}/2 & -1/2 \end{pmatrix},$$

$$\begin{pmatrix} -1 & 0 \\ 0 & 1 \end{pmatrix}, \begin{pmatrix} -1/2 & -\sqrt{3}/2 \\ \sqrt{3}/2 & -1/2 \end{pmatrix}, \text{ and } \begin{pmatrix} -1/2 & \sqrt{3}/2 \\ -\sqrt{3}/2 & -1/2 \end{pmatrix}. \tag{1.10}$$

Surprisingly, this group is isomorphic to the group of the six 3×3 permutation matrices (1.8) of $\mathbf{S}_3$.

   Any matrix group consists of merely invertible matrices (a.k.a. non-singular matrices). The singular matrices (i.e., the matrices with zero determinant) have no inverse.

1.10   SUBGROUPS

An important aspect of a group is formed by its subgroups. For example, the two matrices

$$\begin{pmatrix} 1 & 0 \\ 0 & 1 \end{pmatrix} \text{ and } \begin{pmatrix} -1 & 0 \\ 0 & 1 \end{pmatrix} ,$$

that form a subset of the set (1.10) form a group of their own, isomorphic to $\mathbf{S}_2$, the group of permutations of two objects. We say that this two-element group is a subgroup of the six-element group. We write:

$$\mathbf{S}_2 \subset \mathbf{S}_3 ,$$

which we may read either as "$\mathbf{S}_2$ is a subgroup of $\mathbf{S}_3$" or as "$\mathbf{S}_3$ is a supergroup of $\mathbf{S}_2$."

If $\mathbf{G}$ is finite and a supergroup of $\mathbf{H}$, then the ratio

$$\frac{\text{order}(\mathbf{G})}{\text{order}(\mathbf{H})}$$

is called the index of $\mathbf{H}$ in $\mathbf{G}$. The theorem of Lagrange says that such an index is always an integer. In other words, the order of a subgroup divides the order of its supergroup. This theorem strongly restricts the number of subgroups of a given group. Nevertheless, most groups have a wealth of subgroups. For example, the group $\mathbf{S}_4$ (of order 4! = 24) has 30 different subgroups, whereas $\mathbf{S}_8$ (of order 8! = 40,320) has 151,221 subgroups [18].

We note that Closure($\mathbf{G}_1$, $\mathbf{G}_2$) is a supergroup of both $\mathbf{G}_1$ and $\mathbf{G}_2$.

1.11   YOUNG SUBGROUPS

Symmetric groups have a special class of subgroups called Young subgroups. They take advantage of the notion of "direct product of two groups." We introduce such a product by giving an example of a Young subgroup of $\mathbf{S}_5$. Assume five objects a, b, c, d, and e. There exist a total of 5! = 120 permutations of these objects. However, let us impose a restriction: we only allow permutations that permute a, b, and c among each other and (simultaneously) permute d and e among each other. This allows 3! permutations of three objects, while (independently) allowing 2! permutations of two objects. The allowed permutations form a permutation group of order 3! × 2! = 12. We are allowed to "combine" each element of $\mathbf{S}_3$ with each element of $\mathbf{S}_2$. Therefore, the group is called a direct product of $\mathbf{S}_3$ and $\mathbf{S}_2$ and denoted $\mathbf{S}_3 \times \mathbf{S}_2$. We write

$$\mathbf{S}_3 \times \mathbf{S}_2 \subset \mathbf{S}_5 .$$

The subgroup is based on a particular partition of the number 5:

$$3 + 2 = 5 .$$

In general, we may combine any group $\mathbf{G}_1$ with any group $\mathbf{G}_2$, ..., with any group $\mathbf{G}_m$. Of course, we have

$$\text{order}(\mathbf{G}_1 \times \mathbf{G}_2 \times \cdots \times \mathbf{G}_m) = \text{order}(\mathbf{G}_1) \times \text{order}(\mathbf{G}_2) \times \cdots \times \text{order}(\mathbf{G}_m) \ . \tag{1.11}$$

As an example, we consider the group formed by the set of all 2^{2^n} Boolean functions of n Boolean variables together with the XOR operation (see Sections 1.3 and 1.5). Because in the minterm expansion of a particular function a particular minterm is either present or not, the group is isomorphic to the direct product $\mathbf{S}_2 \times \mathbf{S}_2 \times \cdots \times \mathbf{S}_2$ with 2^n factors, with order

$$\text{order}(\mathbf{S}_2 \times \mathbf{S}_2 \times \cdots \times \mathbf{S}_2) = [\,\text{order}(\mathbf{S}_2)\,]^{2^n} = 2^{2^n} \ .$$

Each of the factors $\mathbf{S}_2$ refers to what was called in Section 1.5 a minterm group.

A Young subgroup [19–21] of the symmetric group $\mathbf{S}_n$ is defined as any subgroup isomorphic to $\mathbf{S}_{n_1} \times \mathbf{S}_{n_2} \times \cdots \times \mathbf{S}_{n_k}$, with $(n_1, n_2, \ldots, n_k)$ a partition of the number n, that is, with

$$n_1 + n_2 + \cdots + n_k = n \ .$$

The order of this Young subgroup is $n_1! n_2! \ldots n_k!$.

For example, the group $\mathbf{S}_4$ has the following Young subgroups:

- one trivial subgroup isomorphic to $\mathbf{S}_4$ (of order $4! = 24$),

- three subgroups isomorphic to $\mathbf{S}_2 \times \mathbf{S}_2$ (each of order $2!2! = 4$),

- four subgroups isomorphic to $\mathbf{S}_1 \times \mathbf{S}_3$ (of order $1!3! = 6$),

- six subgroups isomorphic to $\mathbf{S}_1 \times \mathbf{S}_1 \times \mathbf{S}_2$ (of order $1!1!2! = 2$), and

- one trivial subgroup isomorphic to $\mathbf{S}_1 \times \mathbf{S}_1 \times \mathbf{S}_1 \times \mathbf{S}_1$ (of order 1).

For example,

$$\begin{pmatrix} 1 & 0 & 0 & 0 \\ 0 & 0 & 1 & 0 \\ 0 & 0 & 0 & 1 \\ 0 & 1 & 0 & 0 \end{pmatrix}$$

is a member of an $\mathbf{S}_1 \times \mathbf{S}_3$ subgroup of $\mathbf{S}_4$.

Because $\mathbf{S}_1$ is just the trivial group $\mathbf{1}$ with one element, that is, the identity element i, Young subgroups of the form $\mathbf{S}_1 \times \mathbf{S}_k$ are often simply denoted by $\mathbf{S}_k$. Finally, Young subgroups of the form $\mathbf{S}_k \times \mathbf{S}_k \times \cdot \times \mathbf{S}_k$ (with m factors) will be written as $\mathbf{S}_k^m$.

The theorems in Section 1.8 can be interpreted in terms of Young subgroups: if we define

- $\mathbf{N}$ as the group of all permutations of the n objects,

- $\mathbf{H}$ as the group of all "horizontal permutations" of these objects, and

- **V** as the group of all "vertical permutations,"

then we have that

- **N** is isomorphic to the symmetric group $\mathbf{S}_n$,

- **H** is isomorphic to the Young subgroup $\mathbf{S}_p \times \mathbf{S}_p \times \cdots \times \mathbf{S}_p = \mathbf{S}_p^q$, and

- **V** is isomorphic to the Young subgroup $\mathbf{S}_q \times \mathbf{S}_q \times \cdots \times \mathbf{S}_q = \mathbf{S}_q^p$.

Note that the Young subgroups $\mathbf{S}_p^q$ and $\mathbf{S}_q^p$ are based on two so-called dual partitions of the number n:

$$
\begin{aligned}
n &= p + p + \cdots + p &&(q \text{ terms) and} \\
n &= q + q + \cdots + q &&(p \text{ terms).}
\end{aligned}
\tag{1.12}
$$

Therefore these two subgroups are referred to as dual Young subgroups.

1.12 QUANTUM COMPUTING

We introduce two different kinds of $n \times 1$ matrices, in other words, two kinds of column vectors v:

- a classical vector is a column vector v with all entries equal to 0, except one entry equal to 1; and

- a quantum vector is a column vector v with all entries equal to a complex number v_j, such that the sum $\sum_j v_j \overline{v_j}$ is equal to 1,

where overlining means taking the complex conjugate. It is clear that the former set of vectors is a finite set with n members, whereas the latter set is infinite. Additionally, the former set is a subset of the latter set.

If $n = 2$, then the classical vector $\begin{pmatrix} v_1 \\ v_2 \end{pmatrix}$ is called a bit. If $n = 2^w$, then the classical vector v represents w bits. We consider the classical vectors v of size $n = 2^w$ as the possible inputs for the classical reversible computer. If $v_j = 1$, then the w input bits are given by the binary notation of the number $j - 1$. The output of the computation is given by the column vector $z = Uv$, where U is the permutation matrix of the computer. Automatically z will also be a classical vector. We give an example with the permutation matrix (1.6) and input $(A, B, C) = (1, 1, 0)$, such that

$v_7 = 1$:

$$\begin{pmatrix} 1 & 0 & 0 & 0 & 0 & 0 & 0 & 0 \\ 0 & 1 & 0 & 0 & 0 & 0 & 0 & 0 \\ 0 & 0 & 1 & 0 & 0 & 0 & 0 & 0 \\ 0 & 0 & 0 & 1 & 0 & 0 & 0 & 0 \\ 0 & 0 & 0 & 0 & 1 & 0 & 0 & 0 \\ 0 & 0 & 0 & 0 & 0 & 1 & 0 & 0 \\ 0 & 0 & 0 & 0 & 0 & 0 & 0 & 1 \\ 0 & 0 & 0 & 0 & 0 & 0 & 1 & 0 \end{pmatrix} \begin{pmatrix} 0 \\ 0 \\ 0 \\ 0 \\ 0 \\ 0 \\ 1 \\ 0 \end{pmatrix} = \begin{pmatrix} 0 \\ 0 \\ 0 \\ 0 \\ 0 \\ 0 \\ 0 \\ 1 \end{pmatrix}, \tag{1.13}$$

thus yielding an output vector z with $z_8 = 1$, confirming that input $(A, B, C) = (1, 1, 0)$ gives rise to output $(P, Q, R) = (1, 1, 1)$, as in Table 1.6b.

If $n = 2$, then the quantum vector $\begin{pmatrix} v_1 \\ v_2 \end{pmatrix}$ is called a quantum bit or, for short, a qubit. If $n = 2^w$, then the quantum vector v represents w qubits. If an arbitrary $n \times 1$ quantum vector v is the input of a quantum computer and the $n \times 1$ output $z = Uv$ has to be a quantum vector as well, then it is sufficient and necessary that the transformation matrix U is an $n \times n$ unitary matrix. A unitary matrix is a square matrix U with all entries U_{jk} equal to some complex number, such that the sum $\sum_m U_{jm}\overline{U_{km}}$ equals δ_{jk} for all rows j and k and that the sum $\sum_m U_{mj}\overline{U_{mk}}$ equals δ_{jk} for all columns j and k. Thus each row and each column is a quantum vector, all rows are perpendicular to each other, and so are the columns. In a condensed way, we can say that a matrix U is unitary iff $UU^\dagger$ equals the unit matrix. Here $\dagger$ stands for Hermitian conjugation. An example of a unitary transformation is:

$$\begin{pmatrix} 1/2 & 1/2 & 1/2 & 1/2 \\ 1/2 & i/2 & -1/2 & -i/2 \\ 1/2 & -1/2 & 1/2 & -1/2 \\ 1/2 & -i/2 & -1/2 & i/2 \end{pmatrix} \begin{pmatrix} 0 \\ 1/2 \\ -1/2 \\ (1+i)/2 \end{pmatrix} = \begin{pmatrix} (1+i)/4 \\ 1/2 \\ -(3+i)/4 \\ 0 \end{pmatrix},$$

where i is the imaginary unit.

The $n \times n$ unitary matrices form a group called the unitary group $U(n)$, irrespective of the fact that n equals a power of 2 or not. Its order is infinity. This infinity is not countable, meaning that the set of matrices forms a smooth continuum, just like the real numbers (in contrast to the rational numbers, of which there exist "only" a countable infinity). Continuous groups are often referred to as Lie groups.

Because there exist as many $U(n)$ matrices as there are points in an n^2-dimensional space, we say that the group is of dimension n^2. With a slight abuse of notation, we will write that the order of $U(n)$ equals ∞^{n^2}. For example, there are $\infty^{n^2} = \infty^4$ members U in the group $U(2)$. They thus can be written with the help of $n^2 = 4$ real parameters (θ, φ, ψ, and χ):

$$U = \begin{pmatrix} \cos(\varphi)e^{i(\theta+\psi)} & \sin(\varphi)e^{i(\theta+\chi)} \\ -\sin(\varphi)e^{i(\theta-\chi)} & \cos(\varphi)e^{i(\theta-\psi)} \end{pmatrix}. \tag{1.14}$$

They fill a four-dimensional (curved) space. As an example, we mention the 2×2 discrete Fourier transform, a.k.a. the Hadamard matrix:

$$H = \frac{1}{\sqrt{2}} \begin{pmatrix} 1 & 1 \\ 1 & -1 \end{pmatrix} .$$

It corresponds with the parameter values $\theta = -\pi/2$, $\varphi = \pi/4$, $\psi = \pi/2$, and $\chi = \pi/2$.

In (1.11), we see that the order of a direct product of finite groups equals the product of the orders of the subgroups. We have a similar property for the dimension of continuous groups: the dimension of a direct product of continuous groups equals the sum of the dimensions of its subgroups[3]:

$$\dim(G_1 \times G_2 \times \cdots \times G_m) = \dim(G_1) + \dim(G_2) + \cdots + \dim(G_m) .$$

Thus, the Lie group $U(n_1) \times U(n_2) \times \cdots \times U(n_k)$, subgroup of the Lie group $U(n)$, based on the partition $n = n_1 + n_2 + \cdots + n_k$, has dimension $n_1^2 + n_2^2 + \cdots + n_k^2$. For example

$$\begin{pmatrix} i & 0 & 0 & 0 \\ 0 & 1/4 & (1-3i)/4 & (-2+i)/4 \\ 0 & -(1+3i)/4 & 1/2 & (1+i)/4 \\ 0 & (2+i)/4 & (1-i)/4 & 3/4 \end{pmatrix}$$

is member of a $U(1) \times U(3)$ subgroup of $U(4)$. Wherear $U(4)$ is 16-dimensional, $U(1) \times U(3)$ has only $1^2 + 3^2 = 10$ dimensions.

The finite group $P(n)$ is a subgroup of the Lie group $U(n)$. Table 1.8 gives some orders of $P(n)$ and $U(n)$, for some values $n = 2^w$. We note that any finite group may be considered a zero-dimensional Lie group. Whereas $P(2^w)$ describes classical reversible computers [9], $U(2^w)$ describes quantum computers [22].

1.13   BOTTOM-UP VS. TOP-DOWN

In the previous section, we introduced the infinite group $U(n)$, as a supergroup of the finite group $P(n)$:

$$U(n) \supset P(n) ,$$

---

[3]This is not a great surprise. Indeed, we may write

$$\begin{aligned} \infty^{\dim(G)} &= \text{order}(G) = \text{order}(G_1 \times G_2 \times \cdots \times G_m) \\ &= \text{order}(G_1) \times \text{order}(G_2) \times \cdots \times \text{order}(G_m) \\ &= \infty^{\dim(G_1)} \times \infty^{\dim(G_2)} \times \cdots \times \infty^{\dim(G_m)} \\ &= \infty^{\dim(G_1)+\dim(G_2)+\cdots+\dim(G_m)} \end{aligned}$$

and thus

$$\dim(G) = \dim(G_1) + \dim(G_2) + \cdots + \dim(G_m) .$$

Table 1.8: The number of different (classical) reversible circuits and the number of different quantum circuits, as a function of the number w of (qu)bits

w	Classical	Quantum
1	2	∞^4
2	24	∞^{16}
3	40,320	∞^{64}
4	20,922,789,888,000	∞^{256}

with orders

$$\infty^{n^2} > n! \, .$$

This relationship is depicted in Figure 1.2. Such graph is either read from top to bottom as "U(n) is supergroup of P(n)" or from bottom to top as "P(n) is subgroup of U(n)." In the figure, the group $\mathbf{1}(n)$ is the trivial group consisting of a single matrix, that is, the $n \times n$ unit matrix. This group is isomorphic to $\mathbf{S}_1$ and has order 1.

$$U(n)$$

$$P(n)$$

$$1(n)$$

Figure 1.2: Hierarchy of the Lie group U(n) and the finite groups P(n) and $\mathbf{1}(n)$.

As U(n) is huge compared to P(n), in the next chapters we will look for intermediate groups, simultaneously subgroups of U(n) and supergroups of P(n), thus investigating in detail the relationship between U(n) and P(n), in other words, between quantum computing and classical computing. In next chapters, the simple Figure 1.2 will thus be completed with new, that is, intermediate, groups. We thus look for groups X(n) satisfying

$$P(n) \subset X(n) \subset U(n) \tag{1.15}$$

and thus with order satisfying

$$n! < \text{order}[X(n)] < \infty^{n^2} \, .$$

In Chapter 3 we will proceed from bottom to top. This means we will, step by step, look for ever larger supergroups of the small group P(n) until we arrive at U(n). In Chapter 5 we will proceed from top to bottom. This means we will look for ever smaller subgroups of the large group U(n) until we arrive at P(n).

CHAPTER 2

Bottom

2.1 THE GROUP S_2

Table 2.1 shows the truth table of all reversible gates of width $w = 1$. Each table has one input bit A and one output bit P and consists of $2^w = 2$ rows. Its output column is merely a permutation of its input column. It therefore is a member of the symmetric group S_2.

The group S_2 has two members:

- the unit matrix $\mathtt{I} = \begin{pmatrix} 1 & 0 \\ 0 & 1 \end{pmatrix}$, representing the logic gate called the IDENTITY gate, and

- the matrix $\mathtt{X} = \begin{pmatrix} 0 & 1 \\ 1 & 0 \end{pmatrix}$, representing the logic gate called the NOT gate.

Both gates have one input (qu)bit and one output (qu)bit: $w = 1$. The IDENTITY gate (in fact the absence of any computing) has two different symbols[1]:

$$ -\boxed{\mathtt{I}}- \quad = \quad -\!-\!- \; . $$

The NOT gate also has two symbols:

$$ -\boxed{\mathtt{X}}- \quad = \quad -\!\oplus\!- \; . $$

Table 2.1 gives the corresponding thruth tables.

Table 2.1: Truth table of the two reversible logic circuits of width 1: (a) the IDENTITY gate and (b) the NOT gate

A	P	A	P
0	0	0	1
1	1	1	0
(a)		(b)	

---

[1]All symbols and all schematics in the present book have been edited with the help of the handy special-purpose LaTeX package, called Q-Circuit [23].

2.2   TWO IMPORTANT YOUNG SUBGROUPS OF $\mathbf{S}_{2^w}$

Table 2.2a shows the truth table of an arbitrary reversible gate of width $w = 3$. It has w input bits and w output bits and consists of $2^w = 8$ rows. Its eight output words are merely a permutation of its eight input words 000, 001, ..., 111. It therefore is a member of the symmetric group $\mathbf{S}_n$ with n equal to 2^w.

2.2.1   CONTROLLED CIRCUITS

If n is even, then $\mathbf{S}_n$ has a particular Young subgroup $\mathbf{S}_1^{n/2} \times \mathbf{S}_{n/2}$ (or simply $\mathbf{S}_{n/2}$), represented by $n \times n$ permutation matrices U consisting of two $n/2 \times n/2$ blocks I and U_{22} on the diagonal:

$$U = \begin{pmatrix} I & \mathbf{0} \\ \mathbf{0} & U_{22} \end{pmatrix}, \tag{2.1}$$

where $\mathbf{0}$ denotes the $n/2 \times n/2$ zero matrix, I denotes the $n/2 \times n/2$ unit matrix, and U_{22} is an arbitrary $n/2 \times n/2$ permutation matrix. We recall that the elements of $\mathbf{S}_{2^w}$ describe all reversible circuits acting on w bits. They are represented by the $2^w \times 2^w$ permutation matrices. Among these $2^w!$ matrices, $2^{w-1}!$ are represented by matrices of type (2.1). An example for $w = 3$ is

$$\begin{pmatrix} 1 & 0 & 0 & 0 & 0 & 0 & 0 & 0 \\ 0 & 1 & 0 & 0 & 0 & 0 & 0 & 0 \\ 0 & 0 & 1 & 0 & 0 & 0 & 0 & 0 \\ 0 & 0 & 0 & 1 & 0 & 0 & 0 & 0 \\ 0 & 0 & 0 & 0 & 1 & 0 & 0 & 0 \\ 0 & 0 & 0 & 0 & 0 & 0 & 0 & 1 \\ 0 & 0 & 0 & 0 & 0 & 1 & 0 & 0 \\ 0 & 0 & 0 & 0 & 0 & 0 & 1 & 0 \end{pmatrix}.$$

We can stress its membership of $\mathbf{S}_1^4 \times \mathbf{S}_4$ by writing

$$\begin{pmatrix} 1 & & & & & & & \\ & 1 & & & & & & \\ & & 1 & & & & & \\ & & & 1 & & & & \\ & & & & 1 & 0 & 0 & 0 \\ & & & & 0 & 0 & 0 & 1 \\ & & & & 0 & 1 & 0 & 0 \\ & & & & 0 & 0 & 1 & 0 \end{pmatrix}, \tag{2.2}$$

where empty spaces represent zeroes. In the corresponding truth table, that is, Table 2.2b, output column P_1 is identical to input column A_1. In other words, the reversible circuit does not alter the first bit. Submatrix I tells what happens with the remaining $w - 1$ bits if bit A_1 equals 0: bits

$A_2, A_3, \ldots, A_w$ are not changed. Submatrix U_{22} tells what happens with the remaining $w - 1$ bits if bit A_1 equals 1. The circuit is called a controlled circuit. See Figure 2.1a. We will denote the group of matrices of the kind (2.2) by P$_z(n)$. It is a subgroup of the group P(n) of all $n \times n$ permutation matrices. We call $A_2, A_3, \ldots, A_w$ the controlled bits and A_1 the controlling bit.

Table 2.2: Truth table of three reversible logic circuits of width 3: (a) an arbitrary circuit, (b) a controlled circuit, and (c) a controlled NOT gate

$A_1 A_2 A_3$	$P_1 P_2 P_3$	$A_1 A_2 A_3$	$P_1 P_2 P_3$	$A_1 A_2 A_3$	$P_1 P_2 P_3$
0 0 0	1 1 1	0 0 0	0 0 0	0 0 0	1 0 0
0 0 1	1 1 0	0 0 1	0 0 1	0 0 1	0 0 1
0 1 0	1 0 0	0 1 0	0 1 0	0 1 0	1 1 0
0 1 1	0 0 0	0 1 1	0 1 1	0 1 1	1 1 1
1 0 0	1 0 1	1 0 0	1 0 0	1 0 0	0 0 0
1 0 1	0 1 0	1 0 1	1 1 0	1 0 1	1 0 1
1 1 0	0 0 1	1 1 0	1 1 1	1 1 0	0 1 0
1 1 1	0 1 1	1 1 1	1 0 1	1 1 1	0 1 1
(a)		(b)		(c)	

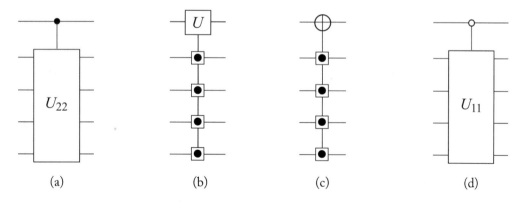

(a)   (b)   (c)   (d)

Figure 2.1: Schematic of four reversible logic circuits of width 5: (a) a controlled circuit: if $A_1 = 1$ then apply U_{22}, (b) a controlled gate: if $f(A_2, A_3, A_4, A_5) = 1$ then apply U, (c) a controlled NOT gate: if $f(A_2, A_3, A_4, A_5) = 1$ then apply $P_1 = \overline{A_1}$, and (d) another controlled circuit: if $A_1 = 0$ then apply U_{11}.

2.2.2   CONTROLLED NOT GATES

We now introduce a second class of reversible logic circuits. They are based on the Young sub-group $\mathbf{S}_2^{n/2} = \mathbf{S}_2^{2^{w-1}}$. We describe these gates, called controlled gates, by means of their relationship between the w outputs $P_1, P_2, \ldots, P_w$ and the w inputs $A_1, A_2, \ldots, A_w$. We have $P_2 = A_2$, $P_3 = A_3$, ..., $P_w = A_w$. The remaining output, i.e., P_1, is controlled by means of some Boolean function f of the $w - 1$ inputs $A_2, A_3, \ldots, A_w$:

- if $f(A_2, A_3, \ldots, A_w) = 0$, then we have $P_1 = A_1$; and

- if, however, $f(A_2, A_3, \ldots, A_w) = 1$, then P_1 is computed from A_1 by applying the gate U (see Figure 2.1b).

In other words: if $f = 0$, then we apply the IDENTITY gate to A_1, otherwise we apply the U gate to A_1. Because, according to Section 2.1, in the classical world, there exist only two possible 1-bit gates U, that is, the IDENTITY gate and the NOT gate, and because a controlled IDENTITY equals an uncontrolled IDENTITY, we only have to consider the case where U is the NOT. The controlled gate then is a controlled NOT. Making use of the XOR function, we can write the rule of the controlled NOT in the following compact way, that is, as a set of w Boolean equations:

$$
\begin{aligned}
P_1 &= f(A_2, A_3, \ldots, A_w) \oplus A_1 \\
P_2 &= A_2 \\
P_3 &= A_3 \\
\ldots &\quad \ldots \\
P_w &= A_w .
\end{aligned}
$$

See Figure 2.1c.

The controlled NOT gates form a group isomorphic to the group of Boolean functions of 2^{w-1} variables. We will denote this group of $2^w \times 2^w$ matrices by $\mathrm{P}_x(2^w)$. It is a subgroup of the group $\mathrm{P}(2^w)$. Its order is $2^{2^{w-1}}$, which is equal to the number of functions f acting on $w - 1$ variables (Section 1.4). We call $A_2, A_3, \ldots, A_w$ the controlling bits and A_1 the controlled bit. We call f the control function. If this function is an AND of the input bits or their negations, then the controlled NOT is called a TOFFOLI gate. For example, in Figure 2.2b, where w equals 5, we have

$$
f(A_2, A_3, A_4, A_5) = A_2 \overline{A_3} A_5 .
$$

Thus, in a TOFFOLI, the control function is a single term and thus a single product of literals. The controlling bits which are not inverted are marked by a black bullet; the controlling bits which are inverted are marked by a white bullet.[2] There are 3^{w-1} different TOFFOLI gates controlling

---

[2]We already encountered a TOFOLLI gate in Section 1.6. There, however, not the first but the last bit is the controlled bit. Bit C is controlled by the controlling bits A and B:

$$
\begin{aligned}
A &\quad\bullet\quad P = A \\
B &\quad\bullet\quad Q = B \\
C &\quad\oplus\quad R = AB \oplus C .
\end{aligned}
$$

the bit A_1. They form a group isomorphic to $\mathbf{C}_3^{w-1}$, where $\mathbf{C}_3$ is the cyclic group of order 3. See Section 1.5.

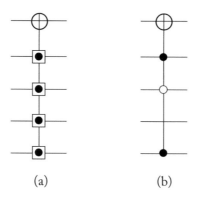

(a)   (b)

Figure 2.2: (a) A controlled NOT gate and (b) a TOFFOLI gate.

Another example of control gate is given by Table 2.2c, where $w = 3$. It has the control function $f(A_2, A_3) = A_2 + \overline{A_3}$. Its permutation matrix is

$$
\begin{pmatrix}
0 & 0 & 0 & 0 & 1 & 0 & 0 & 0 \\
0 & 1 & 0 & 0 & 0 & 0 & 0 & 0 \\
0 & 0 & 0 & 0 & 0 & 0 & 1 & 0 \\
0 & 0 & 0 & 0 & 0 & 0 & 0 & 1 \\
1 & 0 & 0 & 0 & 0 & 0 & 0 & 0 \\
0 & 0 & 0 & 0 & 0 & 1 & 0 & 0 \\
0 & 0 & 1 & 0 & 0 & 0 & 0 & 0 \\
0 & 0 & 0 & 1 & 0 & 0 & 0 & 0
\end{pmatrix} .
$$

We can stress its membership of $\mathbf{S}_2 \times \mathbf{S}_2 \times \mathbf{S}_2 \times \mathbf{S}_2$ by writing

$$
\begin{pmatrix}
0 & & & & 1 & & & \\
& 1 & & & & 0 & & \\
& & 0 & & & & 1 & \\
& & & 0 & & & & 1 \\
1 & & & & 0 & & & \\
& 0 & & & & 1 & & \\
& & 1 & & & & 0 & \\
& & & 1 & & & & 0
\end{pmatrix} ,
\tag{2.3}
$$

a matrix composed of four large $\mathbf{S}_2$ blocks (here one $\begin{pmatrix} 1 & 0 \\ 0 & 1 \end{pmatrix}$ block and three $\begin{pmatrix} 0 & 1 \\ 1 & 0 \end{pmatrix}$ blocks).

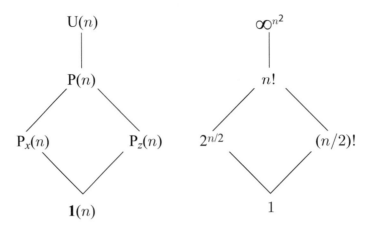

Figure 2.3: Hierarchy of the Lie group $U(n)$ and the finite groups $P(n)$, $P_x(n)$, $P_z(n)$, and $\mathbf{1}(n)$.

2.2.3 CONTROLLED CIRCUITS VS. CONTROLLED GATES

We summarize the present section by Figure 2.3, drawing the group hierarchy: both the group graph and the graph of the corresponding group orders. We recall the duality.

- Members of $P_z(2^w)$ have one controlling bit and $w-1$ controlled bits, whereas members of $P_x(2^w)$ have one controlled bit and $w-1$ controlling bits.

- Matrices from $P_z(n)$ contain 2 blocks from $P(n/2)$, whereas matrices from $P_x(n)$ contain $n/2$ blocks from $P(2)$.

- $P_z(n)$ is isomorphic to $\mathbf{S}_{n/2}$, whereas $P_x(n)$ is isomorphic to $\mathbf{S}_2^{n/2}$.

- $P_z(n)$ has order $(n/2)!$, whereas $P_x(n)$ has order $(2!)^{n/2}$.

We close the present section by introducing yet another subgroup of $P(2^w)$. It consists of the matrices of form

$$U = \begin{pmatrix} U_{11} & \mathbf{0} \\ \mathbf{0} & I \end{pmatrix}.$$

Here again the bits A_2, A_3, ..., and A_w are controlled by the bit A_1. However, here they are affected iff A_1 equals 0. The symbol of this circuit has a white bullet instead of a black one: see Figure 2.1d. We stress that such a circuit can be decomposed into two NOT gates and a positively controlled circuit:

$$\left(\begin{matrix} U_{11} & \\ & I \end{matrix} \right) = \left(\begin{matrix} & I \\ I & \end{matrix} \right) \left(\begin{matrix} I & \\ & U_{11} \end{matrix} \right) \left(\begin{matrix} & I \\ I & \end{matrix} \right) .$$

an equality that also can be written in matrix form:

The matrix $\left(\begin{matrix} & I \\ I & \end{matrix} \right)$ is a matrix of the form (2.3), however with exclusively $\left(\begin{matrix} 0 & 1 \\ 1 & 0 \end{matrix} \right)$ blocks. It is a controlled NOT gate with constant control function $f = 1$. In other words, it is an uncontrolled NOT

We thus are allowed to say that the negatively controlled circuit is decomposed here into

- a member of $P_x(n)$,

- a member of $P_z(n)$, and

- a second member of $P_x(n)$.

2.3   PRIMAL DECOMPOSITION

De Vos and Van Rentergem [24] have demonstrated that it is <u>always</u> possible to decompose an arbitrary classical reversible circuit acting on w bits into three controlled circuits acting on $w - 1$ bits and one controlled NOT gate:

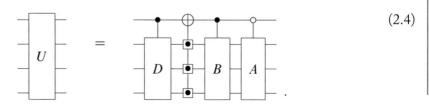

(2.4)

The proof is based on the remarkable theorem of Section 1.8, with $n = 2^w$, $q = 2$ and $p = 2^{w-1}$. In matrix form[3] the theorem reads

$$U = \begin{pmatrix} A & \\ & B \end{pmatrix} C \begin{pmatrix} I & \\ & D \end{pmatrix},$$

where C stands for a matrix of the form (2.3), in other words, a member of $P_x(n)$. Whereas the submatrices A, B, and D correspond to the "horizontal" permutations of Section 1.8, the matrix C corresponds to the "vertical" permutation.

Because of the identity

$$\begin{pmatrix} A & \\ & B \end{pmatrix} = \begin{pmatrix} & I \\ I & \end{pmatrix}\begin{pmatrix} I & \\ & A \end{pmatrix}\begin{pmatrix} & I \\ I & \end{pmatrix}\begin{pmatrix} I & \\ & B \end{pmatrix},$$

we can say that any member U of the group $P(n)$ can be decomposed into three matrices of the group $P_x(n)$ and three matrices of the group $P_z(n)$. This proves that the closure of the groups $P_x(n)$ and $P_z(n)$ equals $P(n)$.

According to Section 1.8, one of the two "horizontal permutations" performs not $q = 2$ but only $q - 1 = 1$ permutations of $p = 2^{w-1}$ objects, such that one of its two permutations is a non-permutation. Hence, this step consists not of two but of only one controlled $(w - 1)$-bit circuit (i.e., controlled circuit D in circuit (2.4)).

The detailed algorithm is as follows [24]. We add to the given truth table (consisting of w input columns A and w output columns P) w extra columns F and w extra columns J. Then, these new columns are filled in, in five steps.

- First, we fill in two columns: column F_1 by copying column A_1 and column J_1 by copying column P_1.

- Then, in the rows satisfying $A_1 = 0$, we fill in $(F_2, F_3, \ldots, F_w) = (J_2, J_3, \ldots, J_w)$ by copying $(A_2, A_3, \ldots, A_w)$ from the same rows.

- Then, in the remaining rows which satisfy $P_1 \oplus P_1' = 1$, we also fill in $(F_2, F_3, \ldots, F_w) = (J_2, J_3, \ldots, J_w)$ by copying $(A_2, A_3, \ldots, A_w)$ from the same rows.

- Then, in the still remaining rows which satisfy $P_1 = 0$, we fill in $(F_2, F_3, \ldots, F_w) = (J_2, J_3, \ldots, J_w)$ by copying $(A_2, A_3, \ldots, A_w)$ from the remaining rows with $P_1 = 1$.

- Finally, in the still remaining rows, we fill in $(F_2, F_3, \ldots, F_w) = (J_2, J_3, \ldots, J_w)$ by copying the still remaining $(A_2, A_3, \ldots, A_w)$ rows.

Above, that is, in the third step, P_1' denotes the bit value in column P_1, but 2^{w-1} rows above the row under consideration.

---

[3]Whereas in a matrix product, subsequent matrices are applied from right to left, in a logic circuit bits are transformed from left to right, according to tradition in engineering. Again it is no surprise that this sometimes (if not often) leads to confusion/errors.

Table 2.3: Expanding the truth table

$A_1A_2A_3$	$P_1P_2P_3$
0 0 0	1 1 1
0 0 1	1 1 0
0 1 0	1 0 0
0 1 1	0 0 0
1 0 0	1 0 1
1 0 1	0 1 0
1 1 0	0 0 1
1 1 1	0 1 1

$A_1A_2A_3$	$F_1F_2F_3$	$J_1J_2J_3$	$P_1P_2P_3$
0 0 0	0 0 0	1 0 0	1 1 1
0 0 1	0 0 1	1 0 1	1 1 0
0 1 0	0 1 0	1 1 0	1 0 0
0 1 1	0 1 1	0 1 1	0 0 0
1 0 0	1	1	1 0 1
1 0 1	1	0	0 1 0
1 1 0	1	0	0 0 1
1 1 1	1	0	0 1 1

Table 2.4: Filling in the expanded truth table according to the primal algorithm

$A_1A_2A_3$	$F_1F_2F_3$	$J_1J_2J_3$	$P_1P_2P_3$
0 0 0	0 0 0	1 0 0	1 1 1
0 0 1	0 0 1	1 0 1	1 1 0
0 1 0	0 1 0	1 1 0	1 0 0
0 1 1	0 1 1	0 1 1	0 0 0
1 0 0	1 **1 1**	1 **1 1**	1 0 1
1 0 1	1 *0 1*	0 0 *1*	0 1 0
1 1 0	1 *1 0*	0 *1 0*	0 0 1
1 1 1	1 **0 0**	**0 0 0**	0 1 1

We illustrate the full procedure for the circuit in Table 2.2a with $w = 3$. Table 2.3 shows the expansion of the table, followed by the first two steps of the procedure. By applying the whole procedure, we obtain Table 2.4. Note that the third step of the procedure is emphasized in italic and the fourth and fifth steps are boldfaced. We see that automatically, between the bits F and the bits J, a controlled NOT is synthesized. Inspection of the F and J columns reveals that here the control function $f(F_2, F_3)$ equals $\overline{F_2} + \overline{F_3}$.

We thus have reduced the synthesis of one 3-bit circuit to the synthesis of three 2-bit circuits and one controlled NOT:

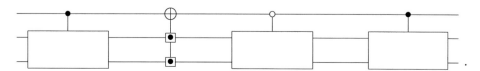

.

In other words, we find a controlled NOT, sandwiched between controlled $(w - 1)$-bit circuits. By applying our procedure to each of the three new circuits, we end up with nine controlled 1-bit circuits and four controlled NOTs:

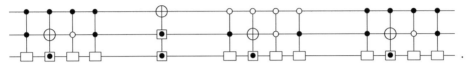

We thus have a total of 13 controlled 1-bit gates. We say that we have a gate cost of 13. In the general problem of synthesizing a w-bit circuit, we find a cascade of $(3^w - 1)/2$ controlled single-bit gates [24]. However, some of the 1-bit circuits turn out to be controlled IDENTITY gates. In our example:

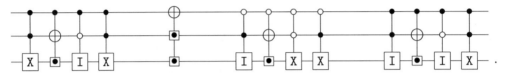

Because a controlled IDENTITY gate equals an uncontrolled IDENTITY gate, we are allowed to delete it from the schematic:

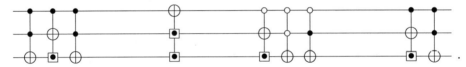

Thus our example synthesis consists of only nine controlled NOT gates,[4] of which at least five are simply TOFFOLI gates.

For an arbitrary reversible circuit acting on w bits, we can equally repeat our decomposition procedure until all controlled circuits are members of $\mathbf{S}_2$, i.e., equal either 1-bit NOT gates or 1-bit IDENTITY gates, with the latter allowed to be ignored. When we apply the approach to each of the $8! = 40{,}320$ circuits of width $w = 3$, we thus obtain a statistical distribution of gate costs, ranging from 0 to $(3^w - 1)/2 = 13$, with about 6.6 average gate cost [12].

2.4   DUAL DECOMPOSITION

Instead of sandwiching a member of $\mathbf{S}_2^8$ between two members of $\mathbf{S}_8^2$, as in (2.4), De Vos and Van Rentergem [12, 25, 26] have demonstrated that we can also work the other way around:

---

[4]The reader may verify that here, in fact, only eight gates are necessary. Post-synthesis optimization indeed can merge two gates:

sandwich a member of $\mathbf{S}_8^2$ between two members of $\mathbf{S}_2^8$:

$$(2.5)$$

This is indeed always possible. Again, a proof is provided by Birkhoff's theorem (Section 1.8), however this time with $n = 2^w$, $p = 2$ and thus $q = 2^{w-1}$. Therefore, we may conclude that a synthesis of an arbitrary circuit of width w may consist of the cascade of

- a first controlled NOT gate,

- two controlled circuits, and

- a second controlled NOT gate.

Note that circuit (2.5) is like (2.4) inside out. Both circuits are called 3-sandwiches [27]. We say that the two circuits are each other's dual. They are indeed based on two dual Young subgroups. These subgroups are based on two dual partitions of the number 2^w:

$$2^w = 2^{w-1} + 2^{w-1}$$
$$= 2 + 2 + \ldots + 2 \qquad (2^{w-1} \text{ terms}),$$

in accordance with (1.12).

The detailed algorithm of the dual decomposition is as follows [12]. Like in the primal decomposition, in the present dual decomposition, we add to the given truth table (consisting of w input columns A and w output columns P) w extra columns F and w extra columns J. We use $F_p(q)$ for the value of bit F_p in the q th row of the truth table and analogously $J_p(q)$ for the value of bit J_p. The columns F_p and J_p are filled in, in three steps.

- First, we fill in $2(w-1)$ columns: columns F_2, F_3, ..., F_w by copying columns A_2, A_3, ..., A_w and columns J_2, J_3, ..., J_w by copying columns P_2, P_3, ..., P_w.

- Then we construct a "coil" of 0's and 1's, starting from bit $F_1(1)$ using a zero label.

- Then we construct a coil starting from the non-filled-in $F_1(j)$ with lowest j, using a zero label, and so on, until all F_1 (and thus also all J_1) are filled in.

Above, a coil consists of a finite number of "windings." Here, a winding is a four-bit sequence $F_1(k) = X$, $F_1(l) = \overline{X}$, $J_1(l) = \overline{X}$, and $J_1(m) = X$, where l results from the condition that the string $F_2(l), F_3(l), \ldots, F_w(l)$ has to be equal to the string $F_2(k), F_3(k), \ldots, F_w(k)$ and where m results from the condition that the string $J_2(m), J_3(m), \ldots, J_w(m)$ has to be equal to the string

$J_2(l), J_3(l), \ldots, J_w(l)$. The number X (with $X \in \{0, 1\}$) we call the "label" of the winding. The first winding of a coil starts at some $F_1(k_1)$; the second winding starts at $F_1(k_2)$, with k_2 equal to m_1, i.e., the row number of the end of the first winding. We continue, winding after winding (all using the same label), until we find an m equal to k_1, i.e., until the coil arrives at $J_1(k_1)$, next to the starting point of the first winding. Then the coil is closed. As all windings of a same coil have the same label, we can refer to it as the label of the coil.

Although the above text might suggest that the algorithm is complicated, it is in fact very straightforward. Tables 2.5 and 2.6 introduce an illustration of this fact by presenting a reversible circuit of width $w = 2$, i.e., the circuit with two inputs (A_1 and A_2) and two outputs ($P_1 = A_2$ and $P_2 = \overline{A_1}$, in this particular case). First, between the columns (A_1, A_2) and (P_1, P_2) of the truth table, we insert the four empty columns (F_1, F_2) and (J_1, J_2). Subsequently, columns F_2 and J_2 are filled in by simply copying columns A_2 and P_2, respectively. This step is shown in Table 2.5 and repeated in Table 2.6. Next comes the tricky part: filling in the columns F_1 and J_1. For the reader's convenience, the boldfaced bits F_1 and J_1 in Table 2.6 are provided with subscripts that tell in which order the bit values are filled in. We start at $F_1(1)$. We may set this bit arbitrarily, but we choose to set it to 0. This starting choice (i.e., the label of the coil) is marked by the subscript 1 in Table 2.6. As a consequence, we automatically can fill in a lot of other bits in columns F_1 and J_1. Indeed, as all computations need to be reversible, $F_1(1) = 0$ automatically leads to $F_1(3) = 1$ (with subscript 2). Then we impose $J_1(3) = F_1(3)$, i.e., $J_1(3) = 1$ (with subscript 3). Again, reversibility requires that $J_1(3) = 1$ infers $J_1(4) = 0$,

Table 2.5: Expanding a 2-bit truth table

$A_1 A_2$	$P_1 P_2$
0 0	0 1
0 1	1 1
1 0	0 0
1 1	1 0

$A_1 A_2$	$F_1 F_2$	$J_1 J_2$	$P_1 P_2$
0 0	0	1	0 1
0 1	1	1	1 1
1 0	0	0	0 0
1 1	1	0	1 0

Table 2.6: Filling in the expanded 2-bit truth table according to the dual algorithm

$A_1 A_2$	$F_1 F_2$	$J_1 J_2$	$P_1 P_2$
0 0	0_1 0	0_8 1	0 1
0 1	1_6 1	1_7 1	1 1
1 0	1_2 0	1_3 0	0 0
1 1	0_5 1	0_4 0	1 0

and so on, until we come back at the starting point $F_1(1)$. In the present example, everything is filled in when the traveling around is closed. So, this synthesis is finished after a single coil (with two windings). This example illustrates that during the application of the algorithm, we "walk in circles," while assigning the bit sequence

$$0, 1, 1, 0, 0, 1, 1, \ldots, 1, 1, 0, 0, 1, 1, 0\,.$$

In case the first coil would be closed before the two columns J_1 and F_1 are completely filled in, the designer just has to start a second coil, and so forth.

We will now apply our procedure for the circuit in Table 2.2a with $w = 3$. Table 2.7 shows the expansion of the table, followed by the first step of the procedure. By applying the full procedure, we obtain Table 2.8. In this example, we have two coils. The former coil (with subscripts from 01 to 12) consists of three windings, whereas the latter coil (with subscripts from 13 to 16) consists of merely one winding.

The above procedure thus yields a decomposition of the logic circuit into three logic circuits, two of which are automatically controlled NOTs. Inspection of Table 2.8 leads to the corresponding control functions $f(A_2, A_3) = A_2 \overline{A_3}$ and $f(J_2, J_3) = J_2$. The third circuit (i.e., the middle circuit) consists of one positively controlled 2-bit circuit and one negatively controlled 2-bit circuit.

We thus have reduced the synthesis of one 3-bit circuit to the synthesis of two 2-bit circuits and two controlled NOTs:

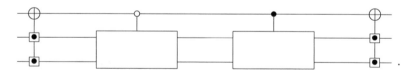

By applying our procedure to each of these two new circuits, we end up with four controlled 1-bit circuits and six controlled NOTs:

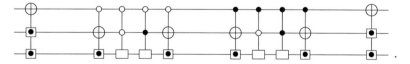

We thus have a total gate cost of 10. In the general problem of synthesizing a w-bit circuit, we find a cascade of $\frac{3}{2} 2^w - 2$ controlled single-bit gates [12, 28]. However, some of the 1-bit circuits may be controlled IDENTITY gates:

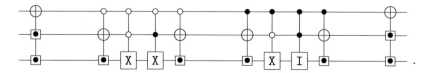

Table 2.7: Expanding the 3-bit truth table

$A_1A_2A_3$	$P_1P_2P_3$
0 0 0	1 1 1
0 0 1	1 1 0
0 1 0	1 0 0
0 1 1	0 0 0
1 0 0	1 0 1
1 0 1	0 1 0
1 1 0	0 0 1
1 1 1	0 1 1

$A_1A_2A_3$	$F_1F_2F_3$	$J_1J_2J_3$	$P_1P_2P_3$
0 0 0	0 0	1 1	1 1 1
0 0 1	0 1	1 0	1 1 0
0 1 0	1 0	0 0	1 0 0
0 1 1	1 1	0 0	0 0 0
1 0 0	0 0	0 1	1 0 1
1 0 1	0 1	1 0	0 1 0
1 1 0	1 0	0 1	0 0 1
1 1 1	1 1	1 1	0 1 1

Table 2.8: Filling in the expanded 3-bit truth table according to the dual algorithm

$A_1A_2A_3$	$F_1F_2F_3$	$J_1J_2J_3$	$P_1P_2P_3$
0 0 0	0_{01} 0 0	0_{12} 1 1	1 1 1
0 0 1	0_{13} 0 1	0_{16} 1 0	1 1 0
0 1 0	1_{06} 1 0	1_{07} 0 0	1 0 0
0 1 1	0_{09} 1 1	0_{08} 0 0	0 0 0
1 0 0	1_{02} 0 0	1_{03} 0 1	1 0 1
1 0 1	1_{14} 0 1	1_{15} 1 0	0 1 0
1 1 0	0_{05} 1 0	0_{04} 0 1	0 0 1
1 1 1	1_{10} 1 1	1_{11} 1 1	0 1 1

Because a controlled IDENTITY gate equals an uncontrolled IDENTITY gate, we are allowed to delete it from the schematic:

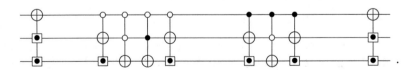

Thus our example synthesis consists of only nine controlled NOT gates, of which at least three are simply TOFFOLI gates. For an arbitrary reversible circuit acting on w bits, we can equally repeat our decomposition procedure until all controlled circuits are members of $\mathbf{S}_2$, in other words, equal either a 1-bit NOT gate or a 1-bit IDENTITY gate, with the latter allowed to be ignored. When we apply the approach to each of the 8! = 40,320 circuits of $w = 3$, we thus obtain a statistical distribution of gate costs [29], ranging from 0 to $\frac{3}{2} 2^w - 2 = 10$, with an average gate cost of about 5.9.

The average result of 5.9 is a lower cost than the average cost of 6.6 in Section 2.3. Also the above maximum cost of $3 \times 2^{w-1} - 2$ is cheaper than the maximum cost of $(3^w - 1)/2$, found in Section 2.3. We note that it is thanks to Theorem 1.2 of Section 1.8 that the primal synthesis method of the previous section needs only a cost of order 3^w. Applying Theorem 1.1 instead of Theorem 1.2 would lead to a cost of order 4^w. In contrast, the dual synthesis method of the present section leads to a cost of order only 2^w, irrespective of which theorem is applied. Choosing Theorem 1.2, as we did above, merely causes one of the two controlled NOTs to have a control function $f(A_2, A_3, \ldots, A_w)$ that satisfies $f(0, 0, \ldots, 0) = 0$. This is explicitly realized in our procedure by starting the first coil with label $F_1(1) = 0$.

2.5   SYNTHESIS EFFICIENCY

According to Sections 2.3 and 2.4, respectively, we need for the primal synthesis of a 3-bit circuit, on average, 6.6 controlled NOTs, whereas for the dual synthesis, on average, only 5.9 controlled NOTs. Below, we will investigate why the dual decomposition is more efficient than the primal one.

Both synthesis methods decompose an arbitrary element of a given group **G** (here the group P(n) of the $n \times n$ permutation matrices) into a product of three building blocks:

- a member g_1 of a subgroup $\mathbf{G}_1$,

- a member g_2 of a subgroup $\mathbf{G}_2$, and

- a member g_3 of a subgroup $\mathbf{G}_3$.

There exist order($\mathbf{G}_1$) order($\mathbf{G}_2$) order($\mathbf{G}_3$) cascades g_1–g_2–g_3. Because some of the products $g_1 g_2 g_3$ can be equal, this yields order($\mathbf{G}_1$) order($\mathbf{G}_2$) order($\mathbf{G}_3$) different products or less different products. As we need synthesis of all order(**G**) elements of **G**, we have the necessary condition

$$\text{order}(\mathbf{G}_1) \, \text{order}(\mathbf{G}_2) \, \text{order}(\mathbf{G}_3) \geq \text{order}(\mathbf{G}) \, . \tag{2.6}$$

An efficient synthesis method has a product order($\mathbf{G}_1$)order($\mathbf{G}_2$)order($\mathbf{G}_3$) that exceeds order(**G**) as little as possible. Then, we indeed synthesize all members of **G** with building blocks that are as simple as possible. We will call the ratio

$$\frac{\text{order}(\mathbf{G}_1) \, \text{order}(\mathbf{G}_2) \, \text{order}(\mathbf{G}_3)}{\text{order}(\mathbf{G})}$$

the synthesis overhead Ω. This number, necessarily equal to or larger than 1, should be as small as possible.

For the primal synthesis strategy above, we have overhead factor

$$\Omega_1 = \frac{(n/2)! \ 2^{n/2} \ [(n/2)!]^2}{n!} \ ,$$

where, as usual, n is the short-hand notation for 2^w. Table 2.9 shows how quickly this quantity grows for increasing numbers of bits.[5] In the case of the dual decomposition, we have an overhead factor of

$$\Omega_2 = \frac{2^{n/2-1} \ [(n/2)!]^2 \ 2^{n/2}}{n!} \ .$$

Table 2.9 shows how this quantity grows very slowly for increasing values of n. Applying Sterling's well-known approximation for the factorial function, we find an approximate value of

$$\frac{1}{4} \ \sqrt{2\pi n} \ .$$

We conclude that the dual decomposition of Section 2.4 indeed is far more efficient than the primal decomposition of Section 2.3, not only for $w = 3$ but for any w.

Table 2.9: Overheads

w	n	Ω_1	Ω_2
1	2	1.00	1.00
2	4	1.33	1.33
3	8	5.49	1.83
4	16	802.01	2.55
5	32	2.28 10^9	3.57

2.6 REFINED SYNTHESIS ALGORITHM

The synthesis based on the partition $2^w = 2^{w-1} + 2^{w-1}$ is discussed in Section 2.3. Because the partition $2^w = 2 + 2 + \cdots + 2$, discussed in Section 2.4, leads to a far more efficient synthesis, it is much more important. Therefore, we will "refine" this algorithm [9, 12].

The algorithm indeed can be "refined" as follows. After applying the dual decomposition, we obtain Figure 2.4b, where the circuit g_1 is simpler than the circuit g, because it fulfills $P_1 = A_1$. By applying the decomposition of g_1 into three circuits, we obtain Figure 2.4c, where the circuit g_2 is again simpler than the circuit g_1, because it fulfills both $P_1 = A_1$ and $P_2 = A_2$. We do so, until we obtain Figure 2.4d, where the circuit g_{w-1} obeys $P_1 = A_1$, $P_2 = A_2$, ..., and

[5]The overheads, of course, are rational numbers. Table 2.9 merely gives floating-point approximations.

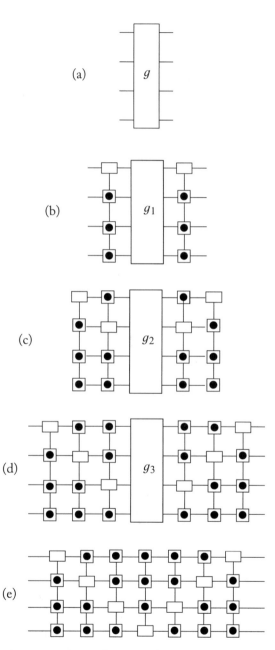

Figure 2.4: Step-by-step decomposition of a reversible logic circuit of width $w = 4$: (a) original logic circuit, (b), (c), and (d) intermediate steps, and (e) refined decomposition.

$P_{w-1} = A_{w-1}$. These properties reveal that g_{w-1} is nothing but a control gate with controlled bit A_w. Therefore, Figure 2.4d is equivalent to Figure 2.4e, such that we have decomposed g into $2w - 1$ controlled NOT gates. This procedure automatically leads us to the refined algorithm.

We add to the given truth table (consisting of w input columns A and w output columns P) not two extra sets of columns, but $2(w - 1)$ sets of columns. We call them A^1, A^2, ..., A^{w-2}, A^{w-1}, P^{w-1}, P^{w-2}, ..., P^2, and P^1. Together they make $2(w - 1)w$ new columns. These are filled in, in the following steps.

- First, we fill in all A^1 columns except column A^1_1 by copying the $w - 1$ corresponding A columns, and analogously, we fill in all P^1 columns except column P^1_1 by copying the $w - 1$ corresponding P columns.

- Then we fill in the two columns A^1_1 and P^1_1 by constructing a coil, starting from bit $A^1_1(1)$, then constructing a new coil, starting at the non-filled-in $A^1_1(j)$ with lowest j, and so on, until all A^1_1 (and thus also all P^1_1) are filled in.

- Then, we fill in all A^2 columns except column A^2_2 by copying the $w - 1$ corresponding A^1 columns, and analogously, we fill in all P^2 columns except column P^2_2 by copying the $w - 1$ corresponding P^1 columns.

- Then we fill in the two columns A^2_2 and P^2_2 by constructing the appropriate number of coils, starting from bit $A^2_2(1)$, until all A^2_2 (and thus also all P^2_2) are filled in.

- We do so, until finally all A^{w-1}_{w-1} (and thus also all P^{w-1}_{w-1}) are filled in. At that moment, we have all $2w^2 2^w$ entries of the extended table.

We end the present section with a historical perspective. The Birkhoff theorem, the basis of the above synthesis procedure, also is the basis of the existence of Clos networks [30], more precisely, rearrangeable (non-blocking) Clos networks [31–33]. The approach has been applied successfully in telephone switching systems and Internet routing [34]. These conventional applications of the Birkhoff theorem are concerned with permutations of wires (or communication channels). Here we apply the theorem not to the w wires, but to the 2^w possible messages.

2.7   EXAMPLES

We illustrate our "refined" procedure by the example circuit in Table 2.2a. By applying the table expansion and the first procedure step, we obtain Table 2.10. By applying the full procedure, we obtain Table 2.11. The second step of the procedure is displayed in italic. This step was already applied in Table 2.8. The third step of the algorithm is underlined.

The above procedure thus yields a decomposition of the logic circuit into five logic circuits (one computing A^1 from A, one computing A^2 from A^1, ..., and one computing P from P^1). All five subcircuits are automatically controlled NOT gates. By merely inspecting Table 2.11,

Table 2.10: Expanded truth table according to the refined algorithm

$A_1A_2A_3$	$A_1^1A_2^1A_3^1$	$A_1^2A_2^2A_3^2$	$P_1^2P_2^2P_3^2$	$P_1^1P_2^1P_3^1$	$P_1P_2P_3$
0 0 0	0 0			1 1	1 1 1
0 0 1	0 1			1 0	1 1 0
0 1 0	1 0			0 0	1 0 0
0 1 1	1 1			0 0	0 0 0
1 0 0	0 0			0 1	1 0 1
1 0 1	0 1			1 0	0 1 0
1 1 0	1 0			0 1	0 0 1
1 1 1	1 1			1 1	0 1 1

Table 2.11: Filling in the expanded truth table according to the refined algorithm

$A_1A_2A_3$	$A_1^1A_2^1A_3^1$	$A_1^2A_2^2A_3^2$	$P_1^2P_2^2P_3^2$	$P_1^1P_2^1P_3^1$	$P_1P_2P_3$
0 0 0	0 0 0	0 0 0	0 0 1	0 1 1	1 1 1
0 0 1	0 0 1	0 0 1	0 0 0	0 1 0	1 1 0
0 1 0	1 1 0	1 0 0	1 0 0	1 0 0	1 0 0
0 1 1	0 1 1	0 1 1	0 1 0	0 0 0	0 0 0
1 0 0	1 0 0	1 1 0	1 1 1	1 0 1	1 0 1
1 0 1	1 0 1	1 1 1	1 1 0	1 1 0	0 1 0
1 1 0	0 1 0	0 1 0	0 1 1	0 0 1	0 0 1
1 1 1	1 1 1	1 0 1	1 0 1	1 1 1	0 1 1

we find their subsequent control functions: $f(A_2, A_3) = A_2\overline{A_3}$, $f(A_1^1, A_3^1) = A_1^1$, $f(A_1^2, A_2^2) = \overline{A_1^2} + A_1^2A_2^2$, $f(P_1^2, P_3^2) = \overline{P_1^2} + P_1^2P_3^2$, and $f(P_2^1, P_3^1) = P_2^1$. The circuit hence looks like

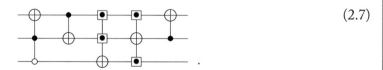

(2.7)

Note that the gate cost of $2w - 1 = 5$ is lower than the gate costs of 9 in both Sections 2.3 and 2.4. Noteworthy is the automatic V-shape of the positions of the five —⊕— symbols in the schematic (2.7).

If we apply the same procedure to each of the 4! = 24 circuits of width $w = 2$, then sometimes one or more of the three control functions equals 0. This means that one or more of the three controlled NOTs is a never-applied NOT gate and is thus in fact absent, such that there is a total of less than three gates. This fact yields a statistical distribution of gate costs ranging from 0 to $L = 2w - 1 = 3$. The average gate cost turns out to be about 1.9. An example for $w = 2$ is the so-called SWAP gate. It swaps two bits: the first output bit equals the second input bit and the second output bit equals the first input bit:

$$P_1 = A_2$$
$$P_2 = A_1 .$$

Its truth table is given in Table 2.12; its circuit symbol is

We find an $L = 3$ decomposition consisting of three controlled NOTs:

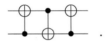

Table 2.12: Truth table of the SWAP function

$A_1 A_2$	$P_1 P_2$
0 0	0 0
0 1	1 0
1 0	0 1
1 1	1 1

When we apply the same procedure to each of the 8! = 40,320 circuits of width $w = 3$, then we obtain a statistical distribution of gate costs ranging from 0 to $L = 2w - 1 = 5$. The average gate cost turns out to be about 4.4. This number is substantially smaller than 6.6, i.e., the average number found with the method of Section 2.3, and even smaller than 5.9, i.e., the average number found with the method of Section 2.4. Table 2.13 gives an overview of the results for $w = 3$ for each of the three synthesis methods.

A circuit like Figure 2.4e is known as a $(2w - 1)$-sandwich [27]. We stress that a gate cost $2w - 1$ is very close to optimal. One can prove [9] that no synthesis method can guarantee better than $2w - 3$. As the controlled NOTs lead to (almost) optimal decompositions, we conclude that they form a natural library for synthesis. We stress that such a library is larger than libraries with merely TOFFOLI gates. We recall that the latter are of the type of Figure 2.2b, i.e., a NOT controlled by some AND function. In the synthesis approach presented here, we fully make use of building blocks of the type of Figure 2.2a, controlled by arbitrary control functions.

Table 2.13: Gate cost of 3-bit circuits, according to the three synthesis methods: minimum cost, average cost, and maximum cost

Method	Min	Ave	Max
Primal	0	6.6	13
Dual	0	5.9	10
Refined	0	4.4	5

Sometimes, for practical purposes, the library of controlled NOT gates is considered too large. Then a library consisting of only TOFFOLI gates is a possibility. In that case, we may proceed as follows: each controlled NOT is decomposed into TOFFOLI gates by merely replacing the control function by its Reed–Muller expansion. From (1.3), we know that such expansion may contain up to $R(w - 1) = 2^{w-1}$ terms. Better results can be obtained by one of the ESOP expansions: only $E(w - 1)$ terms. As an example, we consider the controlled NOT gate with control function $f = A_2 + A_3$, i.e., an OR function:

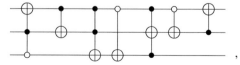

Indeed, its Reed–Muller expansion reads

$$f = A_2 \oplus A_3 \oplus A_2 A_3 ,$$

whereas one of the minimal ESOP expansions is

$$f = A_2 \oplus \overline{A_2}\, A_3 .$$

The minimal ESOP expansion thus often leads to cheaper circuits than the Reed–Muller expansion. For $w = 3$, the Reed–Muller approach can lead to up to $R(2) = 4$ controlled NOTs, whereas the ESOP approach is limited by $E(2) = 2$. The ESOP strategy, however, also has a disadvantage: no efficient algorithm for finding the minimal ESOP exists if $w > 6$. Applying either the Reed–Muller decomposition or a minimal ESOP expansion to the circuit (2.7) yields

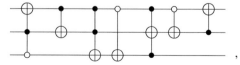

with seven TOFFOLI gates. We conclude that any reversible circuit acting on w bits needs at most

$$(2w - 1)E(w - 1)$$

TOFFOLI gates [35], where the function $E(n)$ satisfies (1.4).

2.8 VARIABLE ORDERING

For each of the three synthesis methods, a lot of fine-tunings exist in order to improve their performance, i.e., to further reduce the resulting gate costs. For example, for the primal synthesis method, we may apply Theorem 1.3 instead of Theorem 1.2. We thus can reposition the controlled circuits, i.e., substitute decomposition

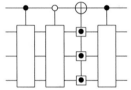

(2.8)

by decomposition

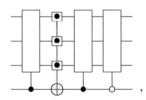

At each of the $w - 1$ steps of the iterative procedure, we can choose the cheaper of the two options.

We can also replace (2.8) by

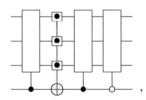

where the bottom bit A_w takes over the role of the top bit A_1. In fact, at each step of the iterative procedure, we can choose which of the remainig bits will act as a controlling/controlled bit.

An interesting optimization of both the primal and the dual synthesis methods consists of applying the following identities:

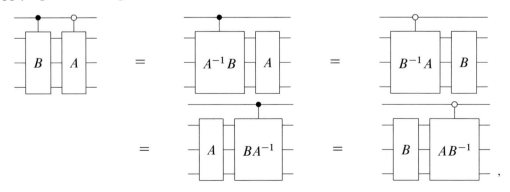

corresponding to the matrix identities

$$\begin{pmatrix} A & \\ & B \end{pmatrix} = \begin{pmatrix} A & \\ & A \end{pmatrix}\begin{pmatrix} I & \\ & A^{-1}B \end{pmatrix} = \begin{pmatrix} B & \\ & B \end{pmatrix}\begin{pmatrix} B^{-1}A & \\ & I \end{pmatrix}$$
$$= \begin{pmatrix} I & \\ & BA^{-1} \end{pmatrix}\begin{pmatrix} A & \\ & A \end{pmatrix} = \begin{pmatrix} AB^{-1} & \\ & I \end{pmatrix}\begin{pmatrix} B & \\ & B \end{pmatrix},$$

where I again stands for the $2^{w-1} \times 2^{w-1}$ unit matrix. These templates do not lower the number of final controlled NOTs, but lower the number of controlling bits.

In Figure 2.4b, we have started the decomposition of Figure 2.4a by applying two controlled NOTs controlling the first bit. Then, in Figure 2.4c, we have applied two control gates controlling the second bit, and so on. There really is no reason to follow this downward order. We may equally well apply any other order. This will lead to $w!$ (usually different) syntheses of the same truth table. For example,

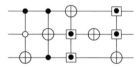

shows the result of applying the upward order to Table 2.2a. In contrast to circuit (2.7), the five —⊕— symbols in the present circuit are not located in V-shape, but in Λ-shape. The gate cost of the two hardware implementations is the same: five units.

For a particular synthesis problem, the $w!$ synthesis solutions may have different hardware cost. On average, however, solving all $(2^w)!$ synthesis problems of width w, with application of the optimal among the $w!$ wire orders instead of a single constant wire order, gives a moderate cost gain. For example, synthesizing all 24 circuits of width 2, trying both wire orderings, yields cascades with average gate cost of about 1.8, only 4% better than the 1.9 result in the previous section. Synthesizing all 40,320 circuits of width 3, trying all six wire orderings, yields cascades with average gate cost of about 3.9, i.e., 11% better than the 4.4 result in the previous section.

Variable ordering and post-synthesis optimizations are automated by RevKit [36, 37], an open-source algebra tool for synthesizing reversible circuits.

CHAPTER 3

Bottom-Up

3.1 THE SQUARE ROOT OF THE NOT

In the bottom-up approach, we take the finite group $\mathbf{S}_n$ (with $n = 2^w$) and one or more extra generators $\{g_1, g_2, \ldots\}$. These generators together with the elements of $\mathbf{S}_n$ generate a new group $\mathbf{G}$, i.e., the closure of $\mathbf{S}_n$ and $\{g_1, g_2, \ldots\}$. It automatically is a supergroup of $\mathbf{S}_n$. Below we give two examples: we add either one (g_1) or two (g_1 and g_2) generators and wonder how much order($\mathbf{G}$) exceeds order($\mathbf{S}_n$) = $n!$.

3.1.1 ONE-(QU)BIT CALCULATIONS

According to Section 1.12, a single qubit is represented by a quantum vector $\begin{pmatrix} a_1 \\ a_2 \end{pmatrix}$, with $a_1 \overline{a_1} + a_2 \overline{a_2} = 1$. If $a_1 = 1$ and $a_2 = 0$, then the input is a classical input bit equal to 0; if $a_1 = 0$ and $a_2 = 1$, then the input is a classical input bit equal to 1. In all other cases, the input is in a quantum-mechanical superposition of 0 and 1:

$$\begin{pmatrix} a_1 \\ a_2 \end{pmatrix} = a_1 \begin{pmatrix} 1 \\ 0 \end{pmatrix} + a_2 \begin{pmatrix} 0 \\ 1 \end{pmatrix}.$$

We introduce, besides the matrices I and X of Section 2.1, the 2×2 unitary matrix

$$V = \frac{1}{2} \begin{pmatrix} 1+i & 1-i \\ 1-i & 1+i \end{pmatrix}.$$

It satisfies $V^2 = X$, with X the NOT gate of Section 2.1:

$$X = \begin{pmatrix} 0 & 1 \\ 1 & 0 \end{pmatrix}.$$

Thus V is the notorious square root of NOT [38–42]:

$$-\boxed{V}-\boxed{V}- \quad = \quad -\boxed{X}- \quad .$$

It plays the role of generator, generating a matrix group of order four with elements

$$I = \begin{pmatrix} 1 & 0 \\ 0 & 1 \end{pmatrix}, \ V = \begin{pmatrix} \zeta & \overline{\zeta} \\ \overline{\zeta} & \zeta \end{pmatrix}, \ X = \begin{pmatrix} 0 & 1 \\ 1 & 0 \end{pmatrix}, \ \text{and } V^\dagger = \begin{pmatrix} \overline{\zeta} & \zeta \\ \zeta & \overline{\zeta} \end{pmatrix},$$

where I stands for the IDENTITY gate, where ζ is the complex number given by

$$\zeta = \frac{1}{2} + i \frac{1}{2},$$

and $\overline{\zeta}$ is its complex conjugate:

$$\overline{\zeta} = \frac{1}{2} - i \frac{1}{2}.$$

The matrix $V^\dagger$ obeys $\left[V^\dagger\right]^2 = X$ and thus is the "other" square root of NOT. Together, the four matrices form a group with respect to the operation of ordinary matrix multiplication, isomorphic to the cyclic group of order 4, i.e., to $\mathbf{C}_4$ (Section 1.5). Indeed, we have $V^1 = V$, $V^2 = X$, $V^3 = V^\dagger$, and $V^4 = I$. We thus have found a group X satisfying the desired property (1.15). Indeed we have

$$P(2) \subset \mathbf{C}_4 \subset U(2),$$

with orders

$$2 < 4 < \infty^4.$$

Any of the four matrices transforms the input state $\begin{pmatrix} a_1 \\ a_2 \end{pmatrix}$ into an output state $\begin{pmatrix} p_1 \\ p_2 \end{pmatrix}$:

$$\begin{pmatrix} p_1 \\ p_2 \end{pmatrix} = \begin{pmatrix} U_{11} & U_{12} \\ U_{21} & U_{22} \end{pmatrix} \begin{pmatrix} a_1 \\ a_2 \end{pmatrix}.$$

Because the matrix U is unitary, $a_1 \overline{a_1} + a_2 \overline{a_2} = 1$ automatically yields $p_1 \overline{p_1} + p_2 \overline{p_2} = 1$. If the input is in a classical state (either $(a_1, a_2) = (1, 0)$ or $(a_1, a_2) = (0, 1)$), then the output is in a quantum superposition. For example,

$$\begin{pmatrix} p_1 \\ p_2 \end{pmatrix} = \begin{pmatrix} \zeta & \overline{\zeta} \\ \overline{\zeta} & \zeta \end{pmatrix} \begin{pmatrix} 1 \\ 0 \end{pmatrix} = \begin{pmatrix} \zeta \\ \overline{\zeta} \end{pmatrix}. \tag{3.1}$$

But, as the output of one circuit may be the input of a subsequent circuit, we have to consider the possibility of (a_1, a_2) being in such a superposition. In fact, we have to consider all possible values of (a_1, a_2) and (p_1, p_2), which may be transformed into one another. These values turn out to be either a column or a row of one of the four matrices. Thus, in total, four and only four states have to be considered: $(1, 0)$, $(0, 1)$, $(\zeta, \overline{\zeta})$, and $(\overline{\zeta}, \zeta)$. Such an object, which may be in four different states, is intermediate to a bit (which can be in only two different states) and a qubit (which may be in as many as ∞^3 different states).

Table 3.1 displays how each of the four matrices acts on the column vector $(a_1 a_2)^T$. The tables constitute the truth tables of the four reversible transformations. Each of these tables expresses a permutation of the four objects $(1, 0)$, $(0, 1)$, $(\zeta, \overline{\zeta})$, and $(\overline{\zeta}, \zeta)$. The four tables replace the two tables in Table 2.1 of Chapter 2. Together they form a permutation group which is a subgroup of the symmetric group $\mathbf{S}_4$. And indeed we have $\mathbf{C}_4 \subset \mathbf{S}_4$.

Table 3.1: The four members of the group with $w = 1$: (a) IDENTITY, (b) square root of NOT, (c) NOT, and (d) "other" square root of NOT

$a_1\,a_2$	$p_1\,p_2$
1 0	1 0
0 1	0 1
$\zeta\,\bar\zeta$	$\zeta\,\bar\zeta$
$\bar\zeta\,\zeta$	$\bar\zeta\,\zeta$
(a)	

$a_1\,a_2$	$p_1\,p_2$
1 0	$\zeta\,\bar\zeta$
0 1	$\bar\zeta\,\zeta$
$\zeta\,\bar\zeta$	0 1
$\bar\zeta\,\zeta$	1 0
(b)	

$a_1\,a_2$	$p_1\,p_2$
1 0	0 1
0 1	1 0
$\zeta\,\bar\zeta$	$\bar\zeta\,\zeta$
$\bar\zeta\,\zeta$	$\zeta\,\bar\zeta$
(c)	

$a_1\,a_2$	$p_1\,p_2$
1 0	$\bar\zeta\,\zeta$
0 1	$\zeta\,\bar\zeta$
$\zeta\,\bar\zeta$	1 0
$\bar\zeta\,\zeta$	0 1
(d)	

3.1.2   TWO AND MULTI-(QU)BIT CALCULATIONS

According to Section 1.12, two qubits exist in a superposition represented by the column vector $(a_1\,a_2\,a_3\,a_4)^T$ with $\sum a_k \overline{a_k} = 1$. The states $(1\,0\,0\,0)^T$, $(0\,1\,0\,0)^T$, $(0\,0\,1\,0)^T$, and $(0\,0\,0\,1)^T$ correspond with the classical bit values (A_1, A_2) equal to $(0, 0)$, $(0, 1)$, $(1, 0)$, and $(1, 1)$, respectively, whereas all other (i.e., superposition) states have no classical equivalent.

The wanted set of 2-qubit circuits should contain all classical reversible 2-bit circuits. Those $4! = 24$ circuits are generated by the two generators g_1 and g_2 (1.5), generating the group P(4) of 4×4 permutation matrices. The set of 2-qubit circuits we investigate [43] also has to comprise the circuit calculating the square root of NOT of qubit # 2. This circuit is represented by the matrix

$$g_3 = \begin{pmatrix} \zeta & \bar\zeta & 0 & 0 \\ \bar\zeta & \zeta & 0 & 0 \\ 0 & 0 & \zeta & \bar\zeta \\ 0 & 0 & \bar\zeta & \zeta \end{pmatrix} = \underline{\quad}\boxed{V}\underline{\quad} . \tag{3.2}$$

Besides a gate V, we also introduce

$$g_4 = \begin{pmatrix} 1 & 0 & 0 & 0 \\ 0 & 1 & 0 & 0 \\ 0 & 0 & \zeta & \bar\zeta \\ 0 & 0 & \bar\zeta & \zeta \end{pmatrix} = \begin{array}{c} \bullet \\ \boxed{V} \end{array} , \tag{3.3}$$

i.e., a gate which may be interpreted either as a "controlled square root of NOT" or as a "square root of controlled NOT."

By adding the generators g_3 and g_4 to the set $\{g_1, g_2\}$, the group P(4) is enlarged[1] to a new group Ω, which surprisingly has infinite order. We can prove this fact [43] by investigating one particular element of Ω, i.e.,

$$y = g_1 g_2 g_4 = \begin{pmatrix} 0 & 0 & \zeta & \bar{\zeta} \\ 0 & 1 & 0 & 0 \\ 0 & 0 & \bar{\zeta} & \zeta \\ 1 & 0 & 0 & 0 \end{pmatrix}.$$

The theory of the Z-transform [43, 44] tells us that the matrix sequence $\{y, y^2, y^3, \ldots\}$ is not periodic. In other words: all matrices y, y^2, y^3, …are different. Therefore, the order of the group element y is infinite. Therefore the order of the group Ω itself is also infinite. One can additionally prove [43] that order(Ω) is countable. Thus we may finally conclude that order(Ω) equals $\aleph_0$. We again have found a group X satisfying the desired property (1.15):

$$P(4) \subset \Omega \subset U(4) ,$$

with orders

$$24 < \aleph_0 < \infty^{16} .$$

We stress here the difference between the two infinities $\aleph_0$ and ∞, the former being the cardinality of the integers, the latter being the cardinality of the reals.

The members M of Ω are U(4) matrices with all 16 entries M_{jk} of the form

$$\frac{1}{2^p} (a_{jk} + i b_{jk}) ,$$

where a_{jk} and b_{jk} are integers, such that $a_{jk} + i b_{jk}$ is a so-called Gaussian integer and p is a non-negative integer. We assume that at least one a_{jk} or one b_{jk} is odd, such that p cannot be lowered by "simplifying fractions." We call p the level of the matrix M. For example, the matrix

$$y^3 = \frac{1}{4} \begin{pmatrix} 2i & 0 & 3-i & 1-i \\ 0 & 4 & 0 & 0 \\ 2 & 0 & -1+i & 3-i \\ 2-2i & 0 & 2 & 2i \end{pmatrix}$$

is a member of Ω with level 2. Whereas in P(4) all matrices have level 0, in Ω, the matrices can have any level. Indeed [45], in U(4), there exist as many as

[1]In fact, generator g_3 is superfluous, as it can be composed from $\{g_1, g_2, g_4\}$:

- 24 matrices of level 0,

- 2,472 matrices of level 1,

- 73,920 matrices of level 2,

- 1,152,000 matrices of level 3,

- and so on.

As circuits of width 2 form a subgroup of the circuits of width w (where w is any integer larger than 2), the same conclusion holds for any width larger than 2. Figure 3.1 shows the group hierarchy. From bottom to top, the respective group orders are:

$$1 < n! < \aleph_0 < \infty^{n^2}.$$

For $n = 2^w$, Table 3.2 gives some numbers.

Table 3.2: The number of different reversible circuits, as a function of the circuit width w: classical, classical plus square roots of NOT, and full quantum

w	Classical	Classical Plus $\sqrt{\text{NOT}}$	Quantum
1	2	4	∞^4
2	24	$\aleph_0$	∞^{16}
3	40,320	$\aleph_0$	∞^{64}
4	20,922,789,888,000	$\aleph_0$	∞^{256}

3.2   MORE ROOTS OF NOT

Inspired by the fact that the X gate is a square root of the I gate (because $X^2 = I$) and V is a square root of X, one can envisage to introduce W, a square root of the V:

$$W = \frac{1}{4} \begin{pmatrix} 2 + \sqrt{2} + i\sqrt{2} & 2 - \sqrt{2} - i\sqrt{2} \\ 2 - \sqrt{2} - i\sqrt{2} & 2 + \sqrt{2} + i\sqrt{2} \end{pmatrix}. \tag{3.4}$$

We indeed have $W^2 = V$. The matrix W generates a cyclic group of order 8, consisting of W, W^2, ..., W^8, where $W^2 = V$, $W^4 = X$, and $W^8 = I$. In circuits with w bits ($w \geq 2$), we can introduce controlled W gates. This results in a group of $2^w \times 2^w$ matrices of order $\aleph_0$, supergroup of $\mathbf{\Omega}(2^w)$.

Of course, one can continue this way indefinitely, introducing higher and higher degrees of roots [46]. Surprising is the fact that (for $w \geq 2$) all resulting groups have the same order, i.e., $\aleph_0$.

Figure 3.1: Hierarchy of the Lie group U(n), the infinite group $\boldsymbol{\Omega}(n)$, and the finite groups P(n) and $\mathbf{1}(n)$.

Indeed, all groups with a countable infinity of elements have the same order. This phenomenon is equivalent to the surprising fact (demonstrated by Cantor) that the number of even numbers 2, 4, 6, 8, …equals the number of natural numbers 1, 2, 3, 4, …

Instead of introducing roots $I^{1/k}$ of the identity gate I with ever increasing integer value of k (i.e., introducing $I^{1/2} = X$, $I^{1/4} = V$, $I^{1/8} = W$, …), below we will relax the restriction that k should be an integer. By letting k be a real number, we will obtain infinite groups with order a non-countable infinity, i.e., ∞^1. The reader may verify that the matrix of the V gate can be written

$$\frac{1}{2}\left(\begin{array}{cc} 1+\omega & 1-\omega \\ 1+\omega & 1-\omega \end{array}\right),\tag{3.5}$$

with ω equal to i, i.e., the square root of -1, i.e., $e^{i\theta}$ with θ equal to $\pi/2$. Analogously the matrix (3.4) of the W gate can be written as (3.5), but with ω equal to the quartic root of -1, i.e., $e^{i\theta}$ with θ equal to $\pi/4$. It will therefore be no surprise that the 1-dimensional infinity of matrices will have the form (3.5), with ω equal to some $e^{i\theta}$ with θ an arbitrary real. We will construct such 1-dimensional sets in the next section.

3.3   NEGATORS

All quantum circuits, acting on a single qubit, are represented by a matrix from U(2). The simplest U(2) matrix is the 2 × 2 unit matrix. We recall that it represents the IDENTITY gate or I gate:

$$\left(\begin{array}{cc} 1 & 0 \\ 0 & 1 \end{array}\right) = \text{I}.$$

The gate is trivial, as it performs no action on the qubit: the output qubit equals the input qubit. The reader can easily verify that, within U(2), the I matrix has a lot of square roots: two diagonal matrices

$$\begin{pmatrix} 1 & 0 \\ 0 & 1 \end{pmatrix} \text{ and } \begin{pmatrix} -1 & 0 \\ 0 & -1 \end{pmatrix},$$

as well as an infinity of matrices:

$$\begin{pmatrix} \cos(\varphi) & \sin(\varphi)e^{i\lambda} \\ \sin(\varphi)e^{-i\lambda} & -\cos(\varphi) \end{pmatrix},$$

where φ and λ are arbitrary real numbers. From this set, we choose two elements:

$$\begin{pmatrix} 0 & 1 \\ 1 & 0 \end{pmatrix} \text{ and } \begin{pmatrix} 1 & 0 \\ 0 & -1 \end{pmatrix}.$$

The former matrix results from the choices $\varphi = \pi/2$ and $\lambda = 0$ and represents the NOT gate or X gate. The latter matrix results from the choice $\varphi = 0$. It is new and is called the Z gate. Whereas the X gate is a classical computer gate, the Z gate is a true quantum gate.[2]

We now introduce generalizations of the X and the Z matrix:

$$\begin{aligned} N(t) &= (1-t)\,\mathrm{I} + t\,\mathrm{X} \\ \Phi(t) &= (1-t)\,\mathrm{I} + t\,\mathrm{Z}, \end{aligned}$$

where t is an interpolation parameter [47]. We can easily prove that these matrices are unitary, iff t is of the form $(1 - e^{i\theta})/2$, resulting in

$$\begin{aligned} N(\theta) &= \tfrac{1}{2}\,(1 + e^{i\theta})\,\mathrm{I} + \tfrac{1}{2}\,(1 - e^{i\theta})\,\mathrm{X} \\ \Phi(\theta) &= \tfrac{1}{2}\,(1 + e^{i\theta})\,\mathrm{I} + \tfrac{1}{2}\,(1 - e^{i\theta})\,\mathrm{Z}. \end{aligned}$$

We thus have constructed two 1-dimensional subgroups of the 4-dimensional group U(2):

$$N(\theta) = \frac{1}{2}\begin{pmatrix} 1 + e^{i\theta} & 1 - e^{i\theta} \\ 1 - e^{i\theta} & 1 + e^{i\theta} \end{pmatrix}$$

$$\text{and } \Phi(\theta) = \begin{pmatrix} 1 & 0 \\ 0 & e^{i\theta} \end{pmatrix}.$$

The gate represented by the matrix $N(\theta)$, we call the NEGATOR gate. It thus constitutes a generalization of the NOT gate. It has the desired form (3.5). The gate represented by the matrix $\Phi(\theta)$, we call the PHASOR gate. We use the following symbols for these quantum gates:

$$-\boxed{N(\theta)}- \quad \text{and} \quad -\boxed{\Phi(\theta)}- \ ,$$

[2]Together, X and I form a group of order 2 (isomorphic to $\mathbf{S_2}$) and so do Z and I. Making all possible products of X, Z, and I generates a group of order 8. It consits of all matrices of the form $\begin{pmatrix} \pm 1 & 0 \\ 0 & \pm 1 \end{pmatrix}$ or $\begin{pmatrix} 0 & \pm 1 \\ \pm 1 & 0 \end{pmatrix}$.

respectively. In the literature [10, 48–50], some of these gates have a particular notation:

$$\begin{aligned}
N(0) &= \text{I}\\
N(\pi/4) &= \text{W}\\
N(\pi/2) &= \text{V}\\
N(\pi) &= \text{X}\\
N(2\pi) &= \text{I}\\[4pt]
\Phi(0) &= \text{I}\\
\Phi(\pi/4) &= \text{T}\\
\Phi(\pi/2) &= \text{S}\\
\Phi(\pi) &= \text{Z}\\
\Phi(2\pi) &= \text{I}\,.
\end{aligned}$$

In particular, the V gate is the square root of NOT, already introduced in Section 3.1, and the W gate is the square root of V, introduced in Section 3.2. Analogously, S is the square root of Z and T is the square root of S.

The subgroup of all $N(\theta)$ matrices, we denote by XU(2); the subgroup of all $\Phi(\theta)$ matrices, we denote by ZU(2). These two Lie subgroups of U(2) have three interesting properties:

Their intersection is minimal, as it equals U(2)'s trivial subgroup $\mathbf{1}(2)$ consisting of merely one 2×2 matrix, i.e., the identity matrix I. Their closure is maximal, as it equals U(2) itself. Indeed, an arbitrary member of U(2), given by (1.14), can be synthesized by cascading merely NEGATORs and PHASORs. Indeed, the reader may verify the following matrix decomposition [47, 51, 52]:

$$\begin{aligned}
U &= \begin{pmatrix} \alpha & 0 \\ 0 & \beta \end{pmatrix} \frac{1}{2} \begin{pmatrix} 1+\gamma & 1-\gamma \\ 1-\gamma & 1+\gamma \end{pmatrix} \begin{pmatrix} 1 & 0 \\ 0 & \delta \end{pmatrix} \\
&= \frac{1}{2} \begin{pmatrix} \alpha(1+\gamma) & \alpha\delta(1-\gamma) \\ \beta(1-\gamma) & \beta\delta(1+\gamma) \end{pmatrix},
\end{aligned} \tag{3.6}$$

where α, β, γ, and δ are complex numbers with unit modulus: either

$$\begin{aligned}
\alpha &= e^{i(\theta+\varphi+\psi)}\\
\beta &= i\,e^{i(\theta+\varphi-\chi)}\\
\gamma &= e^{-2i\varphi}\\
\delta &= -i\,e^{i(-\psi+\chi)}
\end{aligned} \tag{3.7}$$

or

$$\begin{aligned}
\alpha &= e^{i(\theta-\varphi+\psi)}\\
\beta &= -i\,e^{i(\theta-\varphi-\chi)}\\
\gamma &= e^{2i\varphi}\\
\delta &= i\,e^{i(-\psi+\chi)}\,.
\end{aligned} \tag{3.8}$$

Because, moreover, we have the identity

$$
\begin{pmatrix} \alpha & 0 \\ 0 & \beta \end{pmatrix} = \begin{pmatrix} 1 & 0 \\ 0 & \beta \end{pmatrix} \begin{pmatrix} 0 & 1 \\ 1 & 0 \end{pmatrix} \begin{pmatrix} 1 & 0 \\ 0 & \alpha \end{pmatrix} \begin{pmatrix} 0 & 1 \\ 1 & 0 \end{pmatrix}
$$

and because the product $\begin{pmatrix} 0 & 1 \\ 1 & 0 \end{pmatrix} \frac{1}{2} \begin{pmatrix} 1+\gamma & 1-\gamma \\ 1-\gamma & 1+\gamma \end{pmatrix}$ equals $\frac{1}{2} \begin{pmatrix} 1-\gamma & 1+\gamma \\ 1+\gamma & 1-\gamma \end{pmatrix}$, we conclude that U equals the cascade of two NEGATORs and three PHASORs:

$$
\begin{aligned}
U &= \Phi(\theta + \varphi - \chi + \pi/2)\, N(\pi)\, \Phi(\theta + \varphi + \psi)\, N(\pi - 2\varphi)\, \Phi(-\psi + \chi - \pi/2) \\
U &= \Phi(\theta - \varphi - \chi - \pi/2)\, N(\pi)\, \Phi(\theta - \varphi + \psi)\, N(\pi + 2\varphi)\, \Phi(-\psi + \chi + \pi/2)\,.
\end{aligned}
$$

As an example, we take the HADAMARD gate, i.e., the gate performing the Hadamard matrix of Section 1.12. It has the following two decompositions:

$$
\begin{aligned}
H &= \frac{1-i}{\sqrt{2}} \begin{pmatrix} 1 & 0 \\ 0 & i \end{pmatrix} \begin{pmatrix} (1+i)/2 & (1-i)/2 \\ (1-i)/2 & (1+i)/2 \end{pmatrix} \begin{pmatrix} 1 & 0 \\ 0 & i \end{pmatrix} \\
&= e^{-i\pi/4}\, \Phi(\pi/2)\, N(\pi/2)\, \Phi(\pi/2) \\
&= \Phi(\pi/4) N(\pi) \Phi(7\pi/4) N(3\pi/2) \Phi(\pi/2) \\
&= \text{T X ZST XV S}
\end{aligned} \tag{3.9}
$$

and

$$
\begin{aligned}
H &= \frac{1+i}{\sqrt{2}} \begin{pmatrix} 1 & 0 \\ 0 & -i \end{pmatrix} \begin{pmatrix} (1-i)/2 & (1+i)/2 \\ (1+i)/2 & (1-i)/2 \end{pmatrix} \begin{pmatrix} 1 & 0 \\ 0 & -i \end{pmatrix} \\
&= e^{i\pi/4}\, \Phi(-\pi/2)\, N(-\pi/2)\, \Phi(-\pi/2) \\
&= \Phi(7\pi/4) N(\pi) \Phi(\pi/4) N(\pi/2) \Phi(3\pi/2) \\
&= \text{ZST X T V ZS}\,.
\end{aligned} \tag{3.10}
$$

Finally, we note that the two groups XU(2) and ZU(2) are each other's Hadamard conjugate:

$$
\text{ZU(2)} = \text{H XU(2) H} \quad \text{and} \quad \text{XU(2)} = \text{H ZU(2) H}\,. \tag{3.11}
$$

The latter property, which may also be written as

$$
N(\theta) = \text{H}\, \Phi(\theta)\, \text{H}
$$

or

$$
-\boxed{N(\theta)}- \quad = \quad -\boxed{H}-\boxed{\Phi(\theta)}-\boxed{H}- \quad,
$$

leads to an interesting hardware implementation of the NEGATOR: a Mach–Zehnder interferometer including a phase shifter [53].

3.4 NEGATOR CIRCUITS

Two-qubit circuits are represented by matrices from U(4). We may apply either the NEGATOR gate or the PHASOR gate from the previous section to either the first qubit or the second qubit. Here are two examples:

$$\boxed{\Phi(\theta)} \qquad \text{and} \qquad \boxed{N(\theta)} \quad,$$

in other words, a PHASOR acting on the first qubit and a NEGATOR acting on the second qubit, respectively. These circuits are represented by the 4×4 unitary matrices

$$\begin{pmatrix} 1 & 0 & 0 & 0 \\ 0 & 1 & 0 & 0 \\ 0 & 0 & e^{i\theta} & 0 \\ 0 & 0 & 0 & e^{i\theta} \end{pmatrix} \text{ and } \begin{pmatrix} \frac{1}{2}\left(1+e^{i\theta}\right) & \frac{1}{2}\left(1-e^{i\theta}\right) & 0 & 0 \\ \frac{1}{2}\left(1-e^{i\theta}\right) & \frac{1}{2}\left(1+e^{i\theta}\right) & 0 & 0 \\ 0 & 0 & \frac{1}{2}\left(1+e^{i\theta}\right) & \frac{1}{2}\left(1-e^{i\theta}\right) \\ 0 & 0 & \frac{1}{2}\left(1-e^{i\theta}\right) & \frac{1}{2}\left(1+e^{i\theta}\right) \end{pmatrix},$$

respectively.

However, we also introduce the controlled PHASORs and the controlled NEGATORs. Two examples are

$$\boxed{N(\theta)} \qquad \text{and} \qquad \boxed{\Phi(\theta)} \quad,$$

i.e., a positive polarity controlled NEGATOR acting on the first qubit, controlled by the second qubit, and a negative polarity controlled PHASOR acting on the second qubit, controlled by the first qubit, respectively. The former symbol is read as follows: "if the second qubit equals 1, then the NEGATOR acts on the first qubit; if, however, the second qubit equals 0, then the NEGATOR is inactive, i.e., the first qubit undergoes no change." The latter symbol is read as follows: "if the first qubit equals 0, then the PHASOR acts on the second qubit; if, however, the first qubit equals 1, then the PHASOR is inactive, i.e., the second qubit undergoes no change." The matrices representing these circuit examples are:

$$\begin{pmatrix} 1 & 0 & 0 & 0 \\ 0 & \frac{1}{2}\left(1+e^{i\theta}\right) & 0 & \frac{1}{2}\left(1-e^{i\theta}\right) \\ 0 & 0 & 1 & 0 \\ 0 & \frac{1}{2}\left(1-e^{i\theta}\right) & 0 & \frac{1}{2}\left(1+e^{i\theta}\right) \end{pmatrix} \text{ and } \begin{pmatrix} 1 & 0 & 0 & 0 \\ 0 & e^{i\theta} & 0 & 0 \\ 0 & 0 & 1 & 0 \\ 0 & 0 & 0 & 1 \end{pmatrix},$$

respectively.

We now give two examples of a 3-qubit circuit:

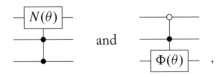

and

i.e., a positive polarity controlled NEGATOR acting on the first qubit and a mixed polarity controlled PHASOR acting on the third qubit. The 8×8 matrices representing these circuit examples are:

$$
\begin{pmatrix}
1 & 0 & 0 & 0 & 0 & 0 & 0 & 0 \\
0 & 1 & 0 & 0 & 0 & 0 & 0 & 0 \\
0 & 0 & 1 & 0 & 0 & 0 & 0 & 0 \\
0 & 0 & 0 & \frac{1}{2}(1+e^{i\theta}) & 0 & 0 & 0 & \frac{1}{2}(1-e^{i\theta}) \\
0 & 0 & 0 & 0 & 1 & 0 & 0 & 0 \\
0 & 0 & 0 & 0 & 0 & 1 & 0 & 0 \\
0 & 0 & 0 & 0 & 0 & 0 & 1 & 0 \\
0 & 0 & 0 & \frac{1}{2}(1-e^{i\theta}) & 0 & 0 & 0 & \frac{1}{2}(1+e^{i\theta})
\end{pmatrix}
\quad\text{and}\quad
\begin{pmatrix}
1 & 0 & 0 & 0 & 0 & 0 & 0 & 0 \\
0 & 1 & 0 & 0 & 0 & 0 & 0 & 0 \\
0 & 0 & 1 & 0 & 0 & 0 & 0 & 0 \\
0 & 0 & 0 & e^{i\theta} & 0 & 0 & 0 & 0 \\
0 & 0 & 0 & 0 & 1 & 0 & 0 & 0 \\
0 & 0 & 0 & 0 & 0 & 1 & 0 & 0 \\
0 & 0 & 0 & 0 & 0 & 0 & 1 & 0 \\
0 & 0 & 0 & 0 & 0 & 0 & 0 & 1
\end{pmatrix},
$$

respectively. We note the following properties:

- the former matrix has all eight row sums and all eight column sums equal to 1 and

- the latter matrix is diagonal and has upper-left entry equal to 1.

Because the multiplication of two square matrices with all line sums equal to 1 automatically yields a third square matrix with all line sums equal to 1, we can easily demonstrate that an arbitrary quantum circuit like

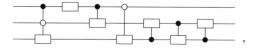

consisting merely of uncontrolled NEGATORs and controlled NEGATORs, is represented by a $2^w \times 2^w$ unitary matrix with all line sums equal to 1. The $n \times n$ unitary matrices with all line sums equal to 1 form a group $XU(n)$, subgroup of $U(n)$. We thus can say that an arbitrary NEGATOR circuit is represented by an $XU(2^w)$ matrix. The converse theorem is also valid: any member X of $XU(2^w)$ can be synthesized by an appropriate string of (un)controlled NEGATORs. The laborious proof [54] of this fact is based on the Hurwitz decomposition [55, 56] of an arbitrary matrix U of the unitary group $U(2^w - 1)$. Such decomposition contains $(2^w - 1) \times (2^w - 1)$ matrices belonging to four different simple categories. This leads to a decomposition of the $2^w \times 2^w$ matrix X into simple factors. It then suffices to prove the theorem for each of the four categories separately.

Because the multiplication of two diagonal square matrices yields a third diagonal square matrix and because the multiplication of two unitary matrices with first entry equal to 1 yields a third unitary matrix with first entry equal to 1, an arbitrary quantum circuit like

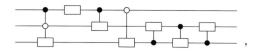

consisting merely of uncontrolled PHASORs and controlled PHASORs, is obviously represented by a $2^w \times 2^w$ unitary diagonal matrix with first entry equal to 1. The $n \times n$ unitary diagonal matrices with upper-left entry equal to 1 form a group $ZU(n)$, subgroup of $U(n)$. We thus can say that an arbitrary PHASOR circuit is represented by a $ZU(2^w)$ matrix. The converse theorem is also valid: any member of $ZU(2^w)$ can be synthesized by an appropriate string of (un)controlled PHASORs (proof being not laborious).

To summarize, the study of NEGATOR and PHASOR circuits leads to the introduction of two subgroups of the unitary group $U(n)$:

- the subgroup $XU(n)$, consisting of all $n \times n$ unitary matrices with all of their $2n$ line sums are equal to 1 and

- the subgroup $ZU(n)$, consisting of all $n \times n$ diagonal unitary matrices with upper-left entry equal to 1.

These subgroups have some properties similar to the three properties of $XU(2)$ and $ZU(2)$, discussed in the previous section. For example, the intersection of $XU(n)$ and $ZU(n)$ is minimal as it is the trivial subgroup $\mathbf{1}(n)$ consisting of a single matrix, i.e., the $n \times n$ unit matrix. However, some properties are not inherited. (See end of Section 3.6.)

3.5   THE GROUP ZU(n)

An arbitrary member Z of $ZU(n)$ is of the form

$$
\begin{pmatrix}
1 & & & & \\
 & e^{i\alpha_2} & & & \\
 & & e^{i\alpha_3} & & \\
 & & & \ddots & \\
 & & & & e^{i\alpha_n}
\end{pmatrix}
= \mathrm{diag}(1, e^{i\alpha_2}, e^{i\alpha_3}, \ldots, e^{i\alpha_n}) .
$$

Because of the identity

$$
\begin{aligned}
\mathrm{diag}(1, a_2, a_3, \ldots, a_{n-1}, a_n) \;=\; & \mathrm{diag}(1, a_2, 1, \ldots, 1, 1)\, \mathrm{diag}(1, 1, a_3, \ldots, 1, 1) \cdots \\
& \mathrm{diag}(1, 1, 1, \ldots, a_{n-1}, 1)\, \mathrm{diag}(1, 1, 1, \ldots, 1, a_n) ,
\end{aligned}
$$

it is clear that the group $ZU(n)$ is isomorphic to the direct product $U(1)^{n-1}$ and thus is an $(n-1)$-dimensional Lie group.

Of course, $ZU(n)$ is a subgroup of $DU(n)$, the n-dimensional group of diagonal unitary matrices, an arbitrary member D of $DU(n)$ being

$$
\mathrm{diag}(e^{i\alpha_1}, e^{i\alpha_2}, e^{i\alpha_3}, \ldots, e^{i\alpha_n}) .
$$

3.6 THE GROUP XU(n)

Any group **G** is isomorphic to a conjugate group $T^{-1}\mathbf{G}T$. Therefore, the group XU(n) is isomorphic to its conjugate

$$T^{-1}\,\mathrm{XU}(n)\,T\;.$$

For the transformation matrix T, we choose the $n \times n$ discrete Fourier transform:

$$F = \frac{1}{\sqrt{n}} \begin{pmatrix} 1 & 1 & 1 & 1 & \cdots & 1 \\ 1 & \omega & \omega^2 & \omega^3 & \cdots & \omega^{n-1} \\ 1 & \omega^2 & \omega^4 & \omega^6 & \cdots & \omega^{2(n-1)} \\ \vdots & & & & & \\ 1 & \omega^{n-1} & \omega^{2(n-1)} & \omega^{3(n-1)} & \cdots & \omega^{(n-1)(n-1)} \end{pmatrix}.$$

Its entries are

$$F_{jk} = \frac{1}{\sqrt{n}}\,\omega^{(j-1)(k-1)}\;,$$

where ω is the primitive n th root of unity: $\omega^n = 1$. For example, for $n = 2$ (and thus $\omega = -1$), the Fourier matrix F equals the Hadamard matrix H of Section 1.12 and, for $n = 4$ (and thus $\omega = i$), we have

$$F = \frac{1}{2} \begin{pmatrix} 1 & 1 & 1 & 1 \\ 1 & i & -1 & -i \\ 1 & -1 & 1 & -1 \\ 1 & -i & -1 & i \end{pmatrix}.$$

For any n, the matrix F is unitary. Hence, if X is an XU(n) matrix, then automatically the matrix $Y = F^{-1}XF$ is a U(n) matrix. The upper-left entry of Y is

$$Y_{11} = \sum_r \sum_s (F^{-1})_{1r}\, X_{rs}\, F_{s1}\;.$$

Because $(F^{-1})_{1r} = F_{s1} = 1/\sqrt{n}$, we obtain

$$Y_{11} = \frac{1}{n} \sum_r \sum_s X_{rs}\;.$$

Because all line sums of X equal unity, its matrix sum $\sum_r \sum_s X_{rs}$ equals n. Hence, we have $Y_{11} = 1$. Because Y is unitary, $Y_{11} = 1$ automatically implies $Y_{1k} = 0$ for all $k \neq 1$ and $Y_{j1} = 0$ for all $j \neq 1$. This means that Y is of the form

$$\begin{pmatrix} 1 & \mathbf{0}_{1\times(n-1)} \\ \mathbf{0}_{(n-1)\times 1} & U \end{pmatrix}, \tag{3.12}$$

where $\mathbf{0}_{a \times b}$ denotes the $a \times b$ zero matrix and U is some $(n-1) \times (n-1)$ unitary matrix. For example, if

$$X = \frac{1}{4} \begin{pmatrix} 3-i & -1+i & 1-i & 1+i \\ 1-i & 3-i & 1+i & -1+i \\ 0 & 0 & 2+2i & 2-2i \\ 2i & 2 & -2i & 2 \end{pmatrix},$$

then

$$Y = \frac{1}{4} \begin{pmatrix} 4 & & & \\ & 1 & 1-3i & -2+i \\ & -1-3i & 2 & 1+i \\ & 2+i & 1-i & 3 \end{pmatrix}.$$

Conversely, it is possible to prove [54] that a matrix of the form

$$F \begin{pmatrix} 1 & \\ & U \end{pmatrix} F^{-1}, \tag{3.13}$$

where U is an arbitrary member of $U(n-1)$, is an $XU(n)$ matrix. The $n \times n$ matrices of the form $\begin{pmatrix} 1 & \\ & U \end{pmatrix}$ create a group isomorphic to the group formed by the matrices U, i.e., the unitary group $U(n-1)$. In Chapter 4, we will call the group consisting of $n \times n$ matrices $\begin{pmatrix} 1 & \\ & U \end{pmatrix}$ the group $aZU(n)$. Because of Equation (3.13), the group $XU(n)$ is isomorphic to $aZU(n)$. Thus we conclude that $XU(n)$ is isomorphic to $U(n-1)$ and therefore is an $(n-1)^2$-dimensional subgroup of the n^2-dimensional group $U(n)$. We note in particular the group chain

$$P(n) \subset XU(n) \subset U(n),$$

where $P(n)$ denotes the finite group of $n \times n$ permutation matrices. We thus found a group satisfying the desired property (1.15). The corresponding group orders in the chain are

$$n! < \infty^{(n-1)^2} < \infty^{n^2}.$$

As all classical reversible circuits (acting on w bits) are represented by a matrix from $P(2^w)$, we may say that the group $XU(2^w)$ represents computers, situated "between" classical computers and full-fledged quantum computers. Such "quantum computers light" cannot perform all quantum computations. However, they can perform some powerful quantum algorithms, such as Grover's notorious search algorithm. Indeed, Grover's diffusion operator $D(n)$ is a unitary matrix with all off-diagonal elements equal $2/n$ and all diagonal elements equal $2/n - 1$ and thus is an $XU(n)$ matrix [57]. For example,

$$D(4) = \frac{1}{2} \begin{pmatrix} -1 & 1 & 1 & 1 \\ 1 & -1 & 1 & 1 \\ 1 & 1 & -1 & 1 \\ 1 & 1 & 1 & -1 \end{pmatrix}$$

is an XU(4) matrix.

If $n > 2$, then XU(n) has more dimensions than ZU(n), as $(n-1)^2$ exceeds $n-1$. See Figure 3.2. Therefore, a conjugation property like (3.11) is not possible. Instead, we have that

$$ \text{ZU}(n) \subset F^{-1}\,\text{XU}(n)\,F \quad \text{and} \quad \text{XU}(n) \supset F\,\text{ZU}(n)\,F^{-1}\,. $$

In particular, the group $F\,\text{ZU}(n)\,F^{-1}$ is the subgroup of XU(n) consisting of all circulant XU(n) matrices. Indeed, if a matrix C equals FDF^{-1}, with D some $n \times n$ diagonal matrix, then the matrix entry

$$ \begin{aligned} C_{jk} &= \sum_r \sum_s F_{jr} D_{rs} (F^{-1})_{sk} = \sum_r F_{jr} D_{rr} (F^{-1})_{rk} \\ &= \frac{1}{n} \sum_r \omega^{(j-1)(r-1)} D_{rr}\, \omega^{-(r-1)(k-1)} = \frac{1}{n} \sum_r \omega^{(j-k)(r-1)} D_{rr} \end{aligned} $$

is a number dependent only on the difference $j - k$. We will denote the group of circulant XU(n) matrices cXU(n). It is isomorphic to ZU(n) and thus has dimension $n - 1$.

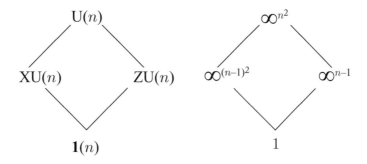

Figure 3.2: Hierarchy of the Lie groups U(n), XU(n), ZU(n), and the finite group $\mathbf{1}(n)$.

3.7 A MATRIX DECOMPOSITION

An important theorem is the ZXZ theorem: any U(n) matrix U can be decomposed as

$$ U = e^{i\alpha}\, Z_1 X Z_2 \,, \tag{3.14} $$

where both Z_1 and Z_2 are ZU(n) matrices and X is an XU(n) matrix. The decomposition is a generalization of (3.6) for n larger than 2. It was conjectured by De Vos and De Baerdemacker [51] and subsequently proved by Idel and Wolf [58]. The proof of the theorem is based on symplectic topology. The proof is not constructive. That means that we know there exists a set $\{\alpha, Z_1, X, Z_2\}$, but (for $n > 2$) we presently lack a general method to find, for a given matrix U, the corresponding number α and matrices Z_1, X, and Z_2.

For some particular (simple) cases, the analytical decomposition is known. For example, if $n = 2$, then the arbitrary unitary matrix, as given by (1.14), has two ZXZ decompositions, given by (3.6)–(3.7)–(3.8). This illustrates that the ZXZ decomposition is not necessarily unique.

If $n > 2$, then we can recur to a numerical iterative algorithm to find (one of) the decomposition(s) with arbitrary precision [51]. For example, the 3×3 unitary matrix

$$\frac{1}{4} \begin{pmatrix} 1 & 1 - 3i & -2 + i \\ -1 - 3i & 2 & 1 + i \\ 2 + i & 1 - i & 3 \end{pmatrix},$$

yields, after only five iterations, a ZXZ decomposition with the following X factor:

$$\begin{pmatrix} -0.2398 + 0.0708\,i & 0.7522 + 0.2432\,i & 0.4337 - 0.3527\,i \\ 0.7113 - 0.3451\,i & 0.4945 + 0.0739\,i & -0.2341 + 0.2649\,i \\ 0.4871 + 0.2742\,i & -0.1564 - 0.3171\,i & 0.7448 + 0.0878\,i \end{pmatrix}.$$

The reader may verify that all six line sums are close to unity. The algorithm is based on a Sinkhorn-like [59] procedure, where the given U(n) matrix is repeatedly left multiplied and right multiplied with a diagonal unitary, until the product approximates an XU(n) matrix sufficiently closely.

We remark that the ZXZ decomposition of a unitary matrix is reminiscent of the HVH decomposition (1.9) of a permutation. As in Section 1.8, we can mention three theorems.

Theorem 3.1   *Each unitary matrix U can be decomposed as*

$$U = D_1 X D_2 , \tag{3.15}$$

where both D_1 and D_2 are diagonal unitary matrices and X is a unit-linesum unitary matrix.

Theorem 3.2   *The ZXZ theorem is a decomposition of the form (3.15), $e^{i\alpha} Z_1$ playing the role of D_1 and Z_2 playing the role of D_2.*

This theorem is slightly more powerful than Theorem 3.1, as the upper-left entry of D_2 is allowed to be equal to 1.

Theorem 3.3   *In the ZXZ decomposition (3.14), the scalar $e^{i\alpha}$ commutes with the matrix $Z_1 X$, yielding a (3.15) decomposition, where Z_1 plays the role of D_1 and $e^{i\alpha} Z_2$ plays the role of D_2.*

Also, this theorem is slightly more powerful than Theorem 3.1, as now the upper-left entry of D_1 is allowed to be equal to 1.

We finally remark that the ZXZ theorem allows us to conclude that the closure of the groups $XU(n)$ and $ZU(n)$ is the group $U(n)$. If n is even, we have the identity

$$\text{diag}(a, a, a, a, a, \ldots, a, a) = P_0 \, \text{diag}(1, a, 1, a, 1, \ldots, 1, a) \, P_0^{-1} \, \text{diag}(1, a, 1, a, 1, \ldots, 1, a) \,,$$
(3.16)

where a is a short-hand notation for $e^{i\alpha}$ and P_0 is the circulant permutation matrix

$$\begin{pmatrix} 0 & 1 & 0 & 0 & \ldots & 0 & 0 \\ 0 & 0 & 1 & 0 & \ldots & 0 & 0 \\ 0 & 0 & 0 & 1 & \ldots & 0 & 0 \\ \vdots & & & & & & \\ 0 & 0 & 0 & 0 & \ldots & 0 & 1 \\ 1 & 0 & 0 & 0 & \ldots & 0 & 0 \end{pmatrix} \,,$$

called the cyclic-shift matrix [9, 60]. If n is odd, then we have analogously

$$\text{diag}(a, a, a, a, a, \ldots, a, a) = P_0 \, \text{diag}(1, a, 1, a, 1, \ldots, a, 1) \, P_0^{-1} \, \text{diag}(1, a, 1, a, 1, \ldots, a, 1) \,.$$

Thus, for all n, the factor $e^{i\alpha}$ in (3.14) can be decomposed into two $XU(n)$ matrices and two $ZU(n)$ matrices. We can conclude that every unitary matrix equals the product of three or less $XU(n)$ members and three or less $ZU(n)$ members. This conclusion is reminiscent of the result in Section 2.3: every permutation matrix is the product of three or less $P_x(n)$ members and three or less $P_z(n)$ members.

3.8   GROUP HIERARCHY

We may summarize above sections by noting the beautiful symmetry

$$ZU(n) \cong U(1)^{n-1} \quad \text{and} \quad XU(n) \cong U(n-1)^1 \,.$$

Analogous to the group hierarchy

$$\mathbf{1}(n) \subset P(n) \subset XU(n) \,,$$

we can construct a hierarchy

$$\mathbf{1}(n) \subset Q(n) \subset ZU(n) \,,$$

by introducing the finite group $Q(n)$. Like $P(n)$ consisting of all $XU(n)$ matrices with integer entries, we define $Q(n)$ as the group consisting of all $ZU(n)$ matrices with exclusively integer entries. Thus, an arbitrary $Q(n)$ matrix is a diagonal matrix with upper-left entry equal to 1 and all other diagonal entries equal ± 1. The group is isomorphic to $\mathbf{S}_2^{n-1}$ and has order 2^{n-1}.

The closure $S(n)$ of the groups $P(n)$ and $Q(n)$ consists of all $n \times n$ matrices with one ± 1 in each row and in each column. The group is called the signed permutation group and has order

$n!2^n$. One can easily prove that any member of S(n) can be decomposed as the product of at most two Q(n) matrices and two P(n) matrices.

We close the present chapter with Figure 3.3, a more detailed and elaborate version of both Figure 1.2 and Figure 3.2. The orders of the groups are displayed in Figure 3.4.

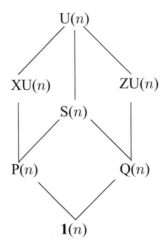

Figure 3.3: Hierarchy of the Lie groups U(n), XU(n), and ZU(n) and the finite groups S(n), P(n), Q(n), and **1**(n).

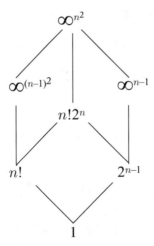

Figure 3.4: Orders of the Lie groups U(n), XU(n), and ZU(n) and the finite groups S(n), P(n), Q(n), and **1**(n).

CHAPTER 4

Top

4.1 PRELIMINARY CIRCUIT DECOMPOSITION

According to (3.14), in other words, according to the decomposition

$$U = e^{i\alpha} Z_1 X Z_2 \,,$$

a quantum schematic (here for $w = 3$ and thus $n = 8$) looks like

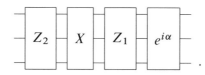

As here n is even, we can apply the identity (3.16). We thus can transform (3.14) into a decomposition containing exclusively XU and ZU matrices:

$$U = P_0 Z_0 P_0^{-1} Z_1' X Z_2 \,,$$

where $Z_0 = \mathrm{diag}(1, e^{i\alpha}, 1, e^{i\alpha}, 1, \dots, 1, e^{i\alpha})$ is a ZU matrix which can be implemented by a single (uncontrolled) PHASOR gate (acting on the w th qubit) and where Z_1' is the product $Z_0 Z_1$:

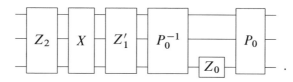

For convenience, we rewrite Equation (3.14) as

$$U = e^{i\alpha_n} L_n X_n R_n \,,$$

where the left matrix L_n and the right matrix R_n are members of ZU(n) and X_n belongs to XU(n). As a member of the $(n-1)^2$-dimensional group XU(n), X_n has the form (3.13):

$$X_n = T_n \begin{pmatrix} 1 & \\ & U_{n-1} \end{pmatrix} T_n^{-1} \,,$$

where U_{n-1} is a member of U($n - 1$) and T_n is the $n \times n$ Fourier matrix F.

Again according to the ZXZ-theorem (3.14), U_{n-1} can be decomposed as

$$e^{i\alpha_{n-1}} \, l_{n-1} x_{n-1} r_{n-1} \, ,$$

a product of a scalar, a ZU$(n-1)$ matrix, an XU$(n-1)$ matrix, and a second ZU$(n-1)$ matrix. We thus obtain for X_n the product $T_n L_{n-1} X_{n-1} R_{n-1} T_n^{-1}$, where

$$L_{n-1} = \begin{pmatrix} 1 & \\ & e^{i\alpha_{n-1}} l_{n-1} \end{pmatrix}, \quad X_{n-1} = \begin{pmatrix} 1 & \\ & x_{n-1} \end{pmatrix}, \quad \text{and} \quad R_{n-1} = \begin{pmatrix} 1 & \\ & r_{n-1} \end{pmatrix}.$$

Hence, we have $U = e^{i\alpha_n} L_n T_n L_{n-1} X_{n-1} R_{n-1} T_n^{-1} R_n$. By applying such decomposition again and again, we find a decomposition

$$e^{i\alpha_n} \, L_n T_n L_{n-1} T_{n-1} L_{n-2} \ldots T_2 L_1 X_1 R_1 T_2^{-1} R_2 \ldots R_{n-2} T_{n-1}^{-1} R_{n-1} T_n^{-1} R_n$$

of an arbitrary member of U(n). As automatically X_1 and R_1 equal the $n \times n$ unit matrix, we thus obtain

$$U = e^{i\alpha_n} \, L_n T_n L_{n-1} T_{n-1} L_{n-2} \ldots T_2 L_1 T_2^{-1} R_2 \ldots R_{n-2} T_{n-1}^{-1} R_{n-1} T_n^{-1} R_n \, , \qquad (4.1)$$

where all n matrices L_j and all $n-1$ matrices R_j belong to the $(n-1)$-dimensional group ZU(n). The $n-1$ matrices T_j are block-diagonal matrices:

$$T_j = \begin{pmatrix} \mathbf{1}_{(n-j)\times(n-j)} & \\ & F_j \end{pmatrix}, \qquad (4.2)$$

where $\mathbf{1}_{(n-j)\times(n-j)}$ is the $(n-j) \times (n-j)$ unit matrix and F_j is the $j \times j$ Fourier matrix. For $w = 2$ (and thus $n = 4$), Equation (4.1) thus looks like the following cascade of six constant matrices, seven ZU circuits, and one overall phase:

R_4	T_4^{-1}	R_3	T_3^{-1}	R_2	T_2^{-1}	L_1	T_2	L_2	T_3	L_3	T_4	L_4	$e^{i\alpha}$
3	0	2	0	1	0	1	0	2	0	3	0	3	1 ,

where the T_j blocks represent the $n-1$ constant matrices

$$T_2 = \begin{pmatrix} 1 & & & \\ & 1 & & \\ & & 1/\sqrt{2} & 1/\sqrt{2} \\ & & 1/\sqrt{2} & -1/\sqrt{2} \end{pmatrix}, \quad T_3 = \begin{pmatrix} 1 & & & \\ & 1/\sqrt{3} & 1/\sqrt{3} & 1/\sqrt{3} \\ & 1/\sqrt{3} & \omega/\sqrt{3} & \omega^2/\sqrt{3} \\ & 1/\sqrt{3} & \omega^2/\sqrt{3} & \omega/\sqrt{3} \end{pmatrix},$$

$$\text{and} \quad T_4 = \begin{pmatrix} 1/2 & 1/2 & 1/2 & 1/2 \\ 1/2 & i/2 & -1/2 & -i/2 \\ 1/2 & -1/2 & 1/2 & -1/2 \\ 1/2 & -i/2 & -1/2 & i/2 \end{pmatrix},$$

with ω equal to the cubic root of unity ($\omega = e^{i\,2\pi/3} = -1/2 + i\,\sqrt{3}/2$). Beneath each of the $4n - 2$ blocks is displayed the number of real parameters of the block. These numbers sum to 16, i.e., exactly n^2, the dimensionality of U(n).

Hence, the synthesis problem of an arbitrary U(2^w) matrix involves, for the given value of w, the synthesis of the $2^w - 1$ circuits T_j. The synthesis of such matrices is not straigthforward [61]. For T_3 above, the following ad hoc decomposition is helpful:

$$T_3 = \begin{pmatrix} 1 & & & \\ & 1 & & \\ & & 1/\sqrt{2} & 1/\sqrt{2} \\ & & 1/\sqrt{2} & -1/\sqrt{2} \end{pmatrix} \begin{pmatrix} 1 & & & \\ & 1/\sqrt{3} & \sqrt{2}/\sqrt{3} & \\ & \sqrt{2}/\sqrt{3} & -1/\sqrt{3} & \\ & & & i \end{pmatrix}$$

$$\begin{pmatrix} 1 & & & \\ & 1 & & \\ & & 1/\sqrt{2} & 1/\sqrt{2} \\ & & 1/\sqrt{2} & -1/\sqrt{2} \end{pmatrix}.$$

However, no general-purpose technique is available for an arbitrary $n \times n$ matrix of the form (4.2) where j satisfies $2 \leq j \leq n - 1$.

The repeated application of the ZXZ theorem thus, in principle, allows the implementation of quantum circuits [61], with the help of 2×2 PHASOR gates and $j \times j$ Fourier-transform circuits with $2 \leq j \leq 2^w$. However, compact and elegant in mathematical form, the ZXZ decomposition is not naturally tailored to qubit-based quantum circuits due to the presence of the $j \times j$ Fourier transforms. Indeed, whenever j is not a power of 2, the Fourier transform acts on a subspace of the total 2^w-dimensional Hilbert space rather than on a subset of the w qubits. In the next section, we will demonstrate a more natural synthesis method, with all j exclusively equal to some 2^a with $1 \leq a \leq w$. It will even turn out that $j = 2$ suffices. Thus only the 2×2 Fourier-transform gate, i.e., the HADAMARD gate, will be necessary.

4.2   PRIMAL DECOMPOSITION

Below we will demonstrate the more natural synthesis method, which respects the qubit structure of the quantum circuit to be synthesized. At each step, the algorithm lowers the matrix size by a factor of 1/2. Thus, instead of matrices from the group sequence U(n), U($n - 1$), U($n - 2$), ..., we will take matrices from U(n), U($n/2$), U($n/4$), ... As a result, the method is not applicable for arbitrary n, but only useful for n equal to some power of 2, i.e., for $n = 2^w$. But this is exactly the n value we need for circuit synthesis.

We recall that an arbitrary member of U(2) can be decomposed as two DU(2) matrices and one XU(2) matrix: see (3.6). We recall that Idel and Wolf [58] proved the generalization for an arbitrary element U of U(n) with arbitrary n:

$$U = D_1 X D_2 \,,$$

where D_1 is an $n \times n$ diagonal unitary matrix, X is an $n \times n$ unitary matrix with all line sums equal to 1, and D_2 is an $n \times n$ diagonal unitary matrix with upper-left entry equal to 1. See Section 3.7. Führ and Rzeszotnik [62] proved another generalization for an arbitrary element U of U(n), however, restricted to even n values:

$$U = \begin{pmatrix} A & 0 \\ 0 & B \end{pmatrix} \frac{1}{2} \begin{pmatrix} I+C & I-C \\ I-C & I+C \end{pmatrix} \begin{pmatrix} I & 0 \\ 0 & D \end{pmatrix}, \qquad (4.3)$$

where A, B, C, and D are matrices from U($n/2$) and I is the $n/2 \times n/2$ unit matrix. We note that, in both generalizations, the number of degrees of freedom is the same in the left-hand side and right-hand side of the equation. In the former case we have

$$n + (n-1)^2 + (n-1) = n^2 ;$$

in the latter case we have

$$2\left(\frac{n}{2}\right)^2 + \left(\frac{n}{2}\right)^2 + \left(\frac{n}{2}\right)^2 = n^2 .$$

We call (4.3) a block-ZXZ decomposition or bZbXbZ decomposition of U.

If n equals 2^w, then the decomposition (4.3) allows for a circuit interpretation. The matrices $\begin{pmatrix} A & \\ & B \end{pmatrix}$ and $\begin{pmatrix} I & \\ & D \end{pmatrix}$ are controlled $(w-1)$-qubit circuits and $\frac{1}{2}\begin{pmatrix} I+C & I-C \\ I-C & I+C \end{pmatrix}$ is an $n \times n$ matrix with all line sums equal 1. We can write

$$\frac{1}{2}\begin{pmatrix} I+C & I-C \\ I-C & I+C \end{pmatrix} = G\begin{pmatrix} I & \\ & C \end{pmatrix} G^{-1},$$

where G is the following $n \times n$ matrix:

$$G = \frac{1}{\sqrt{2}} \begin{pmatrix} I & I \\ I & -I \end{pmatrix}. \qquad (4.4)$$

The (real) matrix G here plays a role similar to the role played by the complex matrix F in Section 3.6. It has a very simple circuit interpretation: a HADAMARD gate acting on the first qubit:

The matrix G can be written as a so-called tensor product: $G = H \otimes I$, where H is the 2×2 Hadamard matrix and I is the $n/2 \times n/2$ unit matrix. We have $G^{-1} = G$.

We conclude that an arbitrary quantum circuit acting on w qubits can be decomposed into two HADAMARD gates and four quantum circuits acting on $w-1$ qubits and controlled by the remaining qubit:

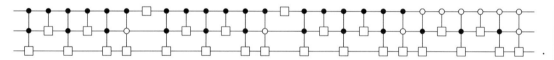

$$(4.5)$$

We now can apply the above decomposition to each of the four circuits A, B, C, and D. By acting so again and again, we finally obtain a decomposition into

- $h = 2(4^{w-1} - 1)/3$ HADAMARD gates, and

- $g = 4^{w-1}$ non-HADAMARD quantum gates acting on a single qubit.

As the former gates have no parameter and each of the latter gates has four parameters, the circuit has $4g = 4^w$ parameters, in accordance with the n^2 degrees of freedom of the given matrix U. We note that all $h + g$ single-qubit gates are controlled gates, with the exception of two HADAMARD gates. For $w = 3$, the circuit looks like

It contains $g + h = \frac{5}{12} 4^w - \frac{2}{3} = 26$ times the symbol —□—.

One might continue the decomposition by decomposing each single-qubit circuit into exclusively NEGATOR gates and PHASOR gates. Indeed, we can rewrite (3.6) as

$$
\begin{pmatrix} 1 & 0 \\ 0 & \beta \end{pmatrix}
\begin{pmatrix} 0 & 1 \\ 1 & 0 \end{pmatrix}
\begin{pmatrix} 1 & 0 \\ 0 & \alpha \end{pmatrix}
\frac{1}{2}\begin{pmatrix} 1-\gamma & 1+\gamma \\ 1+\gamma & 1-\gamma \end{pmatrix}
\begin{pmatrix} 1 & 0 \\ 0 & \delta \end{pmatrix} ,
$$

in other words, a cascade of three PHASOR gates and two NEGATOR gates. One of the latter is simply a NOT gate. In particular for the HADAMARD gate H, we have the decompositions (3.9) and (3.10). We thus end up with a decomposition consisting of $2h + 2g$ NEGATORs and $3h + 3g$ PHASORs. Among the $2h + 2g$ NEGATORs, $h + g$ are NOTs and h are square roots of NOT.

4.3   GROUP STRUCTURE

We recall that the U(n) matrices with all line sums equal to 1 form the subgroup XU(n) of U(n). For even n, the XU(n) matrices of the particular block type

$$
\frac{1}{2}\begin{pmatrix} I+C & I-C \\ I-C & I+C \end{pmatrix} ,
$$

$$(4.6)$$

with $C \in U(n/2)$, form a subgroup $bXU(n)$ of $XU(n)$:

$$U(n) \supset XU(n) \supset bXU(n) \,,$$

with respective dimensions

$$n^2 > (n-1)^2 \geq n^2/4 \qquad \forall n \geq 2 \,.$$

The group structure of $bXU(n)$ follows directly from the group structure of the constituent unitary matrix:

$$\frac{1}{2} \begin{pmatrix} I + C_1 & I - C_1 \\ I - C_1 & I + C_1 \end{pmatrix} \frac{1}{2} \begin{pmatrix} I + C_2 & I - C_2 \\ I - C_2 & I + C_2 \end{pmatrix} = \frac{1}{2} \begin{pmatrix} I + C_1 C_2 & I - C_1 C_2 \\ I - C_1 C_2 & I + C_1 C_2 \end{pmatrix} \,,$$

thus demonstrating the isomorphism $bXU(n) \cong U(n/2)$.

We recall that the diagonal $U(n)$ matrices with upper-left entry equal to 1 form the subgroup $ZU(n)$ of $U(n)$. For even n, the $U(n)$ matrices of the particular block type

$$\begin{pmatrix} I & \\ & D \end{pmatrix} \,, \tag{4.7}$$

with $D \in U(n/2)$, form a group $bZU(n)$, also a subgroup of $U(n)$. The group structure of $bZU(n)$ thus follows trivially from the group structure of $U(n/2)$. Whereas $bXU(n)$ is a subgroup of $XU(n)$, $bZU(n)$ is neither a subgroup nor a supergroup of $ZU(n)$. Whereas $\dim(bXU(n)) \leq \dim(XU(n))$, the dimension of $bZU(n)$, i.e., $n^2/4$, is greater than or equal to the dimension of $ZU(n)$, i.e., $n - 1$.

In Section 3.7, it has been demonstrated that the closure of $XU(n)$ and $ZU(n)$ is the whole group $U(n)$. In other words, any member of $U(n)$ can be written as a product of XU matrices and ZU matrices. Provided n is even, a similar property holds for the block versions of XU and ZU: the closure of $bXU(n)$ and $bZU(n)$ is the whole group $U(n)$. Indeed, with the help of the identity

$$\begin{pmatrix} A & \\ & B \end{pmatrix} = \begin{pmatrix} I & \\ & B \end{pmatrix} \begin{pmatrix} & I \\ I & \end{pmatrix} \begin{pmatrix} I & \\ & A \end{pmatrix} \begin{pmatrix} & I \\ I & \end{pmatrix} \,,$$

we can transform the decomposition (4.3) into a product containing exclusively bXU and bZU matrices, with the particular bXU matrix $\begin{pmatrix} & I \\ I & \end{pmatrix}$, in other words, the block NOT matrix, which represents the (uncontrolled) NOT gate acting on the first qubit; see Section 2.2. This matrix is of the form (4.6) with $C = -I$

Figure 4.1 gives the group structure. The reader is invited to compare with Figure 3.2.

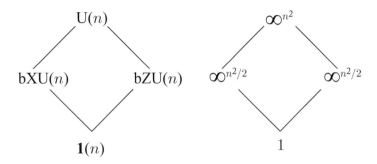

Figure 4.1: Hierarchy of the Lie groups U(n), bXU(n), bZU(n), and the finite group $\mathbf{1}(n)$.

4.4   DUAL DECOMPOSITION

Let U be an arbitrary member of U(n), with n even. We apply the Führ–Rzeszotnik theorem not to U but instead to its conjugate $u = GUG$:

$$u = \begin{pmatrix} a & \\ & b \end{pmatrix} G \begin{pmatrix} I & \\ & c \end{pmatrix} G \begin{pmatrix} I & \\ & d \end{pmatrix}.$$

We decompose the left factor and insert the GG product, equal to the $n \times n$ unit matrix $\begin{pmatrix} I & \\ & I \end{pmatrix}$:

$$U = GuG = G \begin{pmatrix} I & \\ & ba^{-1} \end{pmatrix} GG \begin{pmatrix} a & \\ & a \end{pmatrix} G \begin{pmatrix} I & \\ & c \end{pmatrix} G \begin{pmatrix} I & \\ & d \end{pmatrix} G.$$

Because $G \begin{pmatrix} a & \\ & a \end{pmatrix} G = \begin{pmatrix} a & \\ & a \end{pmatrix}$, we obtain:

$$U = G \begin{pmatrix} I & \\ & ba^{-1} \end{pmatrix} G \begin{pmatrix} a & \\ & ac \end{pmatrix} G \begin{pmatrix} I & \\ & d \end{pmatrix} G,$$

a decomposition of the form

$$U = \frac{1}{2} \begin{pmatrix} I + A' & I - A' \\ I - A' & I + A' \end{pmatrix} \begin{pmatrix} B' & \\ & C' \end{pmatrix} \frac{1}{2} \begin{pmatrix} I + D' & I - D' \\ I - D' & I + D' \end{pmatrix}, \tag{4.8}$$

with

$$A' = ba^{-1}, \; B' = a, \; C' = ac, \text{ and } D' = d. \tag{4.9}$$

We thus obtain a decomposition of the form bXbZbX [63], dual to the Führ–Rzeszotnik decomposition of the form bZbXbZ. We call (4.8) a block-XZX decomposition or bXbZbX decomposition of U. Just like in the bZbXbZ decomposition, the number of degrees of freedom in the bXbZbX decomposition exactly matches the dimension n^2 of the matrix U. For $n = 2^w$, the schematic of the dual decomposition looks like

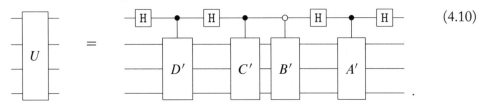
$$\hspace{10cm}(4.10)$$

4.5   DETAILED PROCEDURE

Section 4.2 provides the outline for the primal synthesis of an arbitrary quantum circuit acting on w qubits, given its unitary transformation (i.e., its $2^w \times 2^w$ unitary matrix). However, the synthesis procedure is only complete if, given the matrix U, we are able to actually compute the four matrices A, B, C, and D. Fortunately, in contrast to the Idel–Wolf decomposition (Section 3.7), the Führ–Rzeszotnik decomposition is analytically constructible.

The arbitrary member U of U(2) is given by (1.14). De Vos and De Baerdemacker noticed two different decompositions of this matrix according to (3.6)-(3.7)-(3.8). Führ and Rzeszotnik proved the generalization (4.3) for an arbitrary element

$$U = \begin{pmatrix} U_{11} & U_{12} \\ U_{21} & U_{22} \end{pmatrix}$$

of U(n), for even n values, by introducing the four $n/2 \times n/2$ matrix blocks U_{11}, U_{12}, U_{21}, and U_{22} of U. These four matrices usually are <u>not</u> unitary. Each has a polar decomposition:

$$U_{jk} = P_{jk} V_{jk} \ ,$$

where P_{jk} is a positive semidefinite matrix and V_{jk} is a unitary matrix. For details, see the Appendix.

Just like there are two different expansions in the case $n = 2$, there also exist two decompositions in the case of arbitrary even n. We may choose either [62, 63]

$$\begin{aligned} A &= (P_{11} + i\ P_{12})V_{11} \\ B &= (P_{21} - i\ P_{22})V_{21} \\ C &= V_{11}^{\dagger}(P_{11} - i\ P_{12})^2 V_{11} \\ D &= -i\ V_{11}^{\dagger}V_{12} \end{aligned} \hspace{2cm}(4.11)$$

or [63]

$$\begin{aligned} A &= (P_{11} - i\ P_{12})V_{11} \\ B &= (P_{21} + i\ P_{22})V_{21} \\ C &= V_{11}^{\dagger}(P_{11} + i\ P_{12})^2 V_{11} \\ D &= i\ V_{11}^{\dagger}V_{12} \ . \end{aligned} \hspace{2cm}(4.12)$$

One can verify that both (4.11) and (4.12) satisfy $AA^\dagger = BB^\dagger = CC^\dagger = DD^\dagger = I$, such that A, B, C, and D are all unitary. Finally, one may check that

$$
\begin{aligned}
A(I + C) &= 2U_{11} \\
B(I - C) &= 2U_{21} \\
A(I - C)D &= 2U_{12} \\
B(I + C)D &= 2U_{22},
\end{aligned}
$$

such that (4.3) is fulfilled. Finally, the reader will easily verify that, for $n = 2$, the expressions (4.11) recover the formulae (3.7), and expressions (4.12) recover the formulae (3.8).

We now investigate in more detail the dual decomposition of Section 4.4. Because we have two matrix sets $\{a, b, c, d\}$, we obtain two sets $\{A', B', C', D'\}$:

$$
\begin{aligned}
A' &= (Q_{21} - i\, Q_{22})W_{21}W_{11}^\dagger(Q_{11} - i\, Q_{12}) \\
B' &= (Q_{11} + i\, Q_{12})W_{11} \\
C' &= (Q_{11} - i\, Q_{12})W_{11} \\
D' &= -i\, W_{11}^\dagger W_{12}
\end{aligned}
\tag{4.13}
$$

and

$$
\begin{aligned}
A' &= (Q_{21} + i\, Q_{22})W_{21}W_{11}^\dagger(Q_{11} + i\, Q_{12}) \\
B' &= (Q_{11} - i\, Q_{12})W_{11} \\
C' &= (Q_{11} + i\, Q_{12})W_{11} \\
D' &= i\, W_{11}^\dagger W_{12},
\end{aligned}
$$

respectively. Here, $Q_{jk}W_{jk}$ are the polar decompositions of the four blocks u_{jk} constituting the matrix $u = GUG$.

4.6 EXAMPLES

As an example, we synthesize here the two-qubit circuit realizing the unitary transformation

$$
\frac{1}{12}\begin{pmatrix}
8 & 0 & 4 + 8i & 0 \\
2 + i & 3 - 9i & -2i & -3 - 6i \\
1 - 7i & 6 & -6 + 2i & -3 + 3i \\
3 + 4i & 3 - 3i & 2 - 4i & 9i
\end{pmatrix}.
$$

We perform the algorithm of Section 4.5, applying Hero's iterative method (see Appendix A) for performing the four polar decompositions [64]. Using ten iterations for each Hero decomposition, we thus obtain the following two numerical results:

$$
A = \begin{pmatrix}
0.67 + 0.72i & -0.19 + 0.03i \\
0.18 + 0.06i & 0.80 - 0.57i
\end{pmatrix}, \quad
B = \begin{pmatrix}
-0.33 - 0.64i & 0.50 - 0.47i \\
0.69 + 0.00i & -0.20 - 0.70i
\end{pmatrix},
$$

$$C = \begin{pmatrix} -0.04 - 0.95i & -0.01 - 0.30i \\ -0.07 + 0.29i & 0.25 - 0.92i \end{pmatrix}, \text{ and } D = \begin{pmatrix} 0.87 - 0.43i & -0.15 + 0.20i \\ -0.08 - 0.24i & -0.68 - 0.68i \end{pmatrix}$$

and

$$A = \begin{pmatrix} 0.67 - 0.72i & 0.19 - 0.03i \\ 0.16 + 0.10i & -0.30 - 0.93i \end{pmatrix}, \ B = \begin{pmatrix} 0.50 - 0.52i & 0.50 + 0.47i \\ -0.19 + 0.66i & 0.70 + 0.20i \end{pmatrix},$$

$$C = \begin{pmatrix} -0.04 + 0.95i & -0.07 - 0.29i \\ -0.01 + 0.30i & 0.25 + 0.92i \end{pmatrix}, \text{ and } D = \begin{pmatrix} -0.87 + 0.43i & 0.15 - 0.20i \\ 0.08 + 0.24i & 0.68 + 0.68i \end{pmatrix}.$$

In Chapter 2, we investigated the synthesis of all 24 classical reversible two-bit circuits as well as all 40,320 classical reversible three-bit circuits. We investigated the statistics of the gate cost of the resulting gate cascades. Here, we cannot synthesize all 2-qubit or 3-qubit circuits, for the simple reason there are an infinite number of them. The best we can do is generate, for example, 1,000 random U(4) or U(8) matrices [56, 65] and synthesize the corresponding quantum circuits. However, we can predict that all these circuits will have a same gate cost. Indeed, as explained in Section 4.2, the resulting schematic consists of $\frac{5}{12} 4^w - \frac{2}{3}$ (un)controlled U(2) gates.

Because there exist only two different P(2) blocks, there is one chance out of two such a block equals the IDENTITY block. In contrast, because there exist ∞^4 different U(2) blocks, the probability it equals the IDENTITY block is negligible. Hence, none of the 1,000 circuits will contain a controlled IDENTITY gate that would lead to a cost reduction. Hence, all 2-qubit circuits will have quantum-gate cost equal to 6, and all 3-qubit circuits will have quantum-gate cost equal to 26. Decomposition of each U(2) block into two XU(2) and three ZU(2) blocks (see Section 3.3) does not alter the conclusion: the odds that a NEGATOR gate or a PHASOR gate equals the IDENTITY gate are 1 to ∞[1].

In contrast to the numerical approach in the above first example, we will now perform an analytic decomposition of a second example:

$$U = \begin{pmatrix} 1 & & & \\ & \cos(t) & \sin(t) & \\ & -\sin(t) & \cos(t) & \\ & & & 1 \end{pmatrix},$$

i.e., a typical evolution matrix for spin-spin interaction, often discussed in physics. We have the following four matrix blocks and their polar decompositions[1]:

---

[1]In fact, the presented polar decompositions are only valid if $0 \le t \le \pi/2$ (i.e., if both $c \ge 0$ and $s \ge 0$). However, the reader can easily treat the three other cases.

$$U_{11} = \begin{pmatrix} 1 & 0 \\ 0 & c \end{pmatrix} = \begin{pmatrix} 1 & 0 \\ 0 & c \end{pmatrix} \begin{pmatrix} 1 & 0 \\ 0 & 1 \end{pmatrix}$$

$$U_{12} = \begin{pmatrix} 0 & 0 \\ s & 0 \end{pmatrix} = \begin{pmatrix} 0 & 0 \\ 0 & s \end{pmatrix} \begin{pmatrix} 0 & x \\ 1 & 0 \end{pmatrix}$$

$$U_{21} = \begin{pmatrix} 0 & -s \\ 0 & 0 \end{pmatrix} = \begin{pmatrix} s & 0 \\ 0 & 0 \end{pmatrix} \begin{pmatrix} 0 & -1 \\ y & 0 \end{pmatrix}$$

$$U_{22} = \begin{pmatrix} c & 0 \\ 0 & 1 \end{pmatrix} = \begin{pmatrix} c & 0 \\ 0 & 1 \end{pmatrix} \begin{pmatrix} 1 & 0 \\ 0 & 1 \end{pmatrix},$$

where c and s are short-hand notations for $\cos(t)$ and $\sin(t)$, respectively. Two blocks, that is, U_{12} and U_{21}, are singular and therefore have a polar decomposition which is not unique: both x and y are arbitrary numbers on the unit circle in the complex plane. The resulting two bZbXbZ decompositions of U are:

$$\begin{pmatrix} 1 & & \\ & e & \\ & & -iy \end{pmatrix} \frac{1}{2} \begin{pmatrix} 2 & & \\ 1+1/e^2 & & 1-1/e^2 \\ & 2 & \\ 1-1/e^2 & & 1+1/e^2 \end{pmatrix} \begin{pmatrix} 1 & & \\ & 1 & \\ & & -ix \\ & -i & \end{pmatrix}$$

and

$$\begin{pmatrix} 1 & & \\ & 1/e & \\ & & iy \end{pmatrix} \frac{1}{2} \begin{pmatrix} 2 & & \\ 1+e^2 & & 1-e^2 \\ & 2 & \\ 1-e^2 & & 1+e^2 \end{pmatrix} \begin{pmatrix} 1 & & \\ & 1 & \\ & & ix \\ & i & \end{pmatrix},$$

where e is a short-hand notation for $c + is$. One can easily check that these decompositions are correct iff $xy = -1$. This leaves a 1-dimensional infinitum of decompositions. The fact that some matrices U have an infinity of decompositions is further discussed in the next chapter (i.e., Chapter 5).

4.7 SYNTHESIS EFFICIENCY

Both synthesis methods (i.e., the primal one and the dual one) decompose an arbitrary element of a given Lie group $\mathbf{G}$ (here the group $U(n)$ of the $n \times n$ unitary matrices) into a product of three building blocks: a member of a subgroup $\mathbf{G}_1$, a member of a subgroup $\mathbf{G}_2$, and a member of a subgroup $\mathbf{G}_3$. By merely substituting in (2.6) "order" by "dimension" and replacing products by sums and division by subtraction, we obtain the overhead formula for Lie groups:

$$\dim(\mathbf{G}_1) + \dim(\mathbf{G}_2) + \dim(\mathbf{G}_3) - \dim(\mathbf{G}),$$

an integer number equal to or greater than 0. For both of our two synthesis strategies, the formula gives a zero overhead, expressing that both methods are ideally efficient.

4.8   FURTHER SYNTHESIS

For the synthesis of classical reversible circuits, after the primal synthesis (Section 2.3) and the dual synthesis (Section 2.4), Section 2.6 presented a "refined synthesis." Is such a third synthesis method also possible for quantum circuits? Unfortunately, the answer is "no."

Indeed, for classical computing, a controlled 2×2 matrix can only be either a controlled IDENTITY gate or a controlled NOT gate. As a controlled IDENTITY gate, in fact, it is performing no action at all; it can be deleted from any circuit. Hence, for classical computing, a controlled 2×2 matrix is a controlled NOT gate. For quantum computing, a controlled 2×2 gate can be any of the ∞^4 matrices of U(2). Therefore, a controlled U(2) gate, acting on the first qubit, is represented by a matrix like (2.3), however, with the 2^{w-1} blocks, either $\begin{pmatrix} 1 & 0 \\ 0 & 1 \end{pmatrix}$ or $\begin{pmatrix} 0 & 1 \\ 1 & 0 \end{pmatrix}$ replaced by blocks from U(2). For $w = 3$, this looks like

$$
\begin{pmatrix}
a_1 & & & & b_1 & & & \\
& a_2 & & & & b_2 & & \\
& & a_3 & & & & b_3 & \\
& & & a_4 & & & & b_4 \\
c_1 & & & & d_1 & & & \\
& c_2 & & & & d_2 & & \\
& & c_3 & & & & d_3 & \\
& & & c_4 & & & & d_4
\end{pmatrix},
\tag{4.14}
$$

where each block $\begin{pmatrix} a_j & b_j \\ c_j & d_j \end{pmatrix}$ is some U(2) matrix. Hence, each controlled U(2) has 2^{w-1} times four degrees of freedom, thus 2^{w+1} degrees of freedom. A cascade of $2w - 1$ controlled U(2) gates thus has a total of $(2w - 1)2^{w+1}$ parameters. With $n = 2^w$, this means $2n[2\log_2(n) - 1]$ degrees of freedom, i.e., too few to synthesize an arbitrary member of the n^2-dimensional group U(n).

We thus can conclude that a "refined synthesis" method is impossible for quantum computing. Fortunately, such a method is not necessary, as both the "primal synthesis" method and the "dual synthesis" method are already optimal in the quantum case.

We finaly remark that, both in quantum circuit (4.5) and in quantum circuit (4.10), we can change the order of the blocks, as mentioned in Section 2.8 for classical circuits. We also can vary the order of the quantum wires. And, finally, we can reduce the number of controlling qubits.

4.9   AN EXTRA DECOMPOSITION

We close the present chapter by drawing the hierarchy of the unitary group U(n) for even n: Figure 4.2. We note the extra subgroup of U(n), casually introduced in Section 3.6: the group aZU(n) of $n \times n$ unitary matrices with upper-left entry equal to unity.[2] In other words, the aZU matrices are the matrices of form (3.12).

From top to bottom of the graph, we recognize:

- the group U(n) of dimension n^2,

- the groups XU(n) and aZU(n), both of dimension $(n - 1)^2$, each other's Fourier conjugate,

- the groups bXU(n) and bZU(n), both of dimension $(n/2)^2$, each other's Hadamard conjugate,

- the groups cXU(n) and ZU(n), both of dimension $n - 1$, each other's Fourier conjugate, and

- the group $\mathbf{1}(n)$ of zero dimension.

Thanks to the groups bXU(n) and bZU(n), we have both the primal bZbXbZ decomposition (Section 4.2) and the dual bXbZbX decomposition (Section 4.4). Thanks to the groups XU(n) and ZU(n), we have the ZXZ decomposition (Section 3.7). It therefore is no surprise that the groups cXU(n) and aZU(n) also lead to a matrix decomposition. Indeed, we can apply a trick similar to the one in Section 4.4: we apply the ZXZ decomposition (3.14) not to an arbitrary matrix U of U(n), but to its conjugate $u = F^{-1}UF$:

$$u = e^{i\alpha} Z_1 X Z_2 \, .$$

We insert two $F^{-1}F$ products and thus find

$$
\begin{aligned}
U = FuF^{-1} &= Fe^{i\alpha}Z_1 \, F^{-1}F \, X \, F^{-1}F \, Z_2F^{-1} \\
&= e^{i\alpha} \, (FZ_1F^{-1})(FXF^{-1})(FZ_2F^{-1}) = e^{i\alpha} \, C_1 A \, C_2 \, ,
\end{aligned}
$$

i.e., a decomposition into a scalar $e^{i\alpha}$, two matrices (i.e., C_1 and C_2) from cXU(n) and one matrix (i.e., A) from aZU(n). Because the proof of the ZXZ decomposition is not constructive, the CAC decomposition theorem also is not constructive. We have to use numerical methods to find approximations for the number α and the matrices C_1, A, and C_2.

---

[2]Here, the prescript a in the expression aZU has no meaning, in contrast to b in bXU and bZU meaning "block" and c in cXU meaning "circulant." It is chosen for the sake of symmetry.

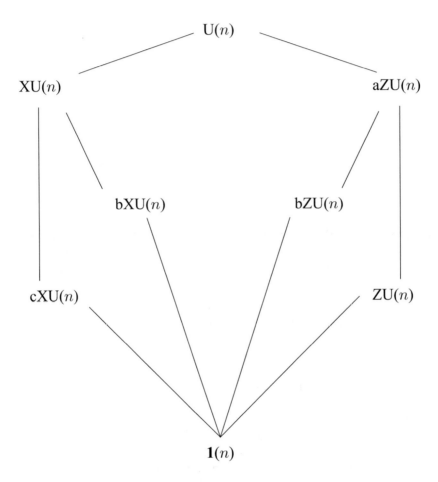

Figure 4.2: Hierarchy of the Lie groups U(n), XU(n), bXU(n), cXU(n), ZU(n), aZU(n), and bZU(n) and the finite group $\mathbf{1}(n)$.

CHAPTER 5

Top-Down

5.1 TOP VS. BOTTOM

In Chapter 2, we developed two synthesis methods for an arbitrary classical reversible circuit: a primal method and a dual method, the former based on the decomposition (2.4) of an arbitrary permutation matrix, the latter based on decomposition (2.5). In Chapter 5, we developed two synthesis methods for an arbitrary quantum circuit: a primal method and a dual method, the former based on the decomposition (4.5) of an arbitrary unitary matrix, the latter based on decomposition (4.10).

Decomposition (2.4), that is,

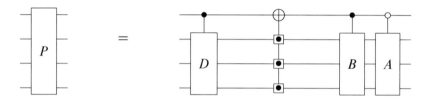

and decomposition (4.5), that is,

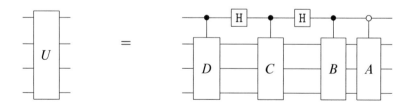

have been deduced and proved completely independently. And so have decomposition (2.5), that is,

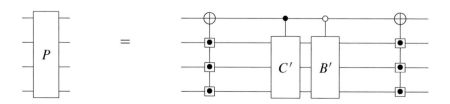

and decomposition (4.10), that is,

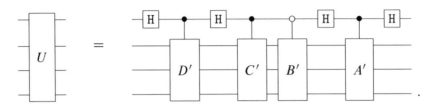

The question arises whether (2.4) can be deduced from (4.5) and whether (2.5) can be deduced from (4.10), as a permutation matrix is a special case of a unitary matrix. This question will be addressed in the present chapter.

5.2   LIGHT MATRICES

As a special case of the two synthesis methods of Chapter 4, we consider for U a permutation matrix. Such a choice is particularly interesting, as a $2^w \times 2^w$ permutation matrix represents a classical reversible computation on w bits, i.e., the subject of Chapter 2. For $w = 2$, we investigate the example

$$U = \begin{pmatrix} 0 & 1 & 0 & 0 \\ 0 & 0 & 0 & 1 \\ 1 & 0 & 0 & 0 \\ 0 & 0 & 1 & 0 \end{pmatrix}. \tag{5.1}$$

Its four constitutive blocks and their polar decompositions are

$$U_{11} = \begin{pmatrix} 0 & 1 \\ 0 & 0 \end{pmatrix} = \begin{pmatrix} 1 & 0 \\ 0 & 0 \end{pmatrix} \begin{pmatrix} 0 & 1 \\ x & 0 \end{pmatrix}$$

$$U_{12} = \begin{pmatrix} 0 & 0 \\ 0 & 1 \end{pmatrix} = \begin{pmatrix} 0 & 0 \\ 0 & 1 \end{pmatrix} \begin{pmatrix} y & 0 \\ 0 & 1 \end{pmatrix}$$

$$U_{21} = \begin{pmatrix} 1 & 0 \\ 0 & 0 \end{pmatrix} = \begin{pmatrix} 1 & 0 \\ 0 & 0 \end{pmatrix} \begin{pmatrix} 1 & 0 \\ 0 & z \end{pmatrix}$$

$$U_{22} = \begin{pmatrix} 0 & 0 \\ 1 & 0 \end{pmatrix} = \begin{pmatrix} 0 & 0 \\ 0 & 1 \end{pmatrix} \begin{pmatrix} 0 & w \\ 1 & 0 \end{pmatrix},$$

where x, y, z, and w are arbitrary unit-modulus numbers. Thus, in this example, none of the four polar decompositions $U_{jk} = P_{jk} V_{jk}$ is unique. If, in particular, we choose $x = w = -i$ and $y = z = i$, then we find, by means of (4.11), a primal bZbXbZ decomposition of U consisting exclusively of permutation matrices:

$$\begin{pmatrix} 0 & 1 & & \\ 1 & 0 & & \\ & & 1 & 0 \\ & & 0 & 1 \end{pmatrix} \begin{pmatrix} 0 & & 1 & \\ & 1 & & 0 \\ 1 & & 0 & \\ & 0 & & 1 \end{pmatrix} \begin{pmatrix} 1 & 0 & & \\ 0 & 1 & & \\ & & 0 & 1 \\ & & 1 & 0 \end{pmatrix}.$$

In the next section, we will demonstrate that such decomposition is possible for any $n \times n$ permutation matrix (provided n is even).

The above example leads us to a deeper analysis of sparse unitary matrices.

Definition 5.1   Let M be an $m \times m$ matrix with, in each row and each column, maximum one non-zero entry. We call such a sparse matrix "light." Let μ be the number of non-zero entries of M. We call μ the weight of M. We have $0 \le \mu \le m$. If $\mu = m$, then M is regular; if $\mu < m$, then M is singular.

The reader will easily understand the proof for the following two lemmas.

Lemma 5.2   *Let PU (with P a positive-semidefinite matrix and U a unitary matrix) be the polar decomposition of an $m \times m$ light matrix M. Then P is a diagonal matrix and U is a complex permutation matrix. If μ, the weight of M, equals m, then U is unique; otherwise, we have an $(m - \mu)$-dimensional infinity of choices for U.*

Lemma 5.3   *If P is a diagonal matrix and U is a complex permutation matrix, then $U^\dagger PU$ is a diagonal matrix, with the same entries as P, in a permuted order.*

We now combine these two lemmas. Assume that the $n \times n$ matrix U consists of four $n/2 \times n/2$ blocks, such that the two blocks U_{11} and U_{12} are light. Then, by virtue of Lemma 5.2, the positive-semidefinite matrices P_{11} and P_{12} are diagonal. Therefore $P_{11} - iP_{12}$ is diagonal and so is $(P_{11} - iP_{12})^2$. By virtue of Lemma 5.2 again, the matrix V_{11} is a complex permutation matrix. Finally, because of Lemma 5.3, the matrix $C = V_{11}^\dagger(P_{11} - iP_{12})^2 V_{11}$ in (4.11) is diagonal and so are $I + C$ and $I - C$. As a result, for $n = 2^w$, the matrix $G\left(\begin{smallmatrix} I & \\ & C \end{smallmatrix} \right)G = \frac{1}{2}\left(\begin{smallmatrix} I+C & I-C \\ I-C & I+C \end{smallmatrix} \right)$ is of a form like (4.14) with all $\left(\begin{smallmatrix} a & b \\ c & d \end{smallmatrix} \right)$ of NEGATOR type and thus represents a cascade of 2^{w-1} NEGATOR gates acting on the first qubit and controlled by the $w - 1$ other qubits:

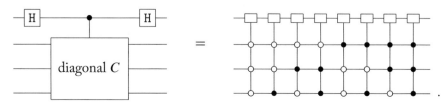

Miraculously, the two HADAMARDs have disappeared from the synthesis. We now are in a position to discuss the case of U being an $n \times n$ permutation matrix, representing a classical reversible computation.

5.3 PRIMAL DECOMPOSITION

First, we will prove that $\frac{1}{2} \left(\begin{smallmatrix} I+C & I-C \\ I-C & I+C \end{smallmatrix} \right)$ is a structured permutation matrix. If U is an $n \times n$ permutation matrix, then both $n/2 \times n/2$ blocks U_{11} and U_{12} are light, the sum of their weights μ_{11} and μ_{12} being equal to $n/2$. The matrices P_{11} and P_{12} are diagonal, with entries equal to 0 or 1, with the special feature that, wherever there is a zero entry in P_{11}, the matrix P_{12} has a 1 on the same row, and vice versa. The matrix $P_{11} - iP_{12}$ thus is diagonal, with all diagonal entries either equal to 1 or to $-i$. Hence, the matrix $(P_{11} - iP_{12})^2$ is diagonal, with all diagonal entries either equal to 1 or to -1, and so is matrix C according to (4.11). Hence, the matrices $I + C$ and $I - C$ are diagonal with entries either 0 or 2. As a result, for $n = 2^w$, the matrix $G\left(\begin{smallmatrix} I & \\ & C \end{smallmatrix} \right)G = \frac{1}{2} \left(\begin{smallmatrix} I+C & I-C \\ I-C & I+C \end{smallmatrix} \right)$ represents a cascade of 1-qubit IDENTITY and NOT gates acting on the first qubit and controlled by the $w - 1$ other qubits. Thus, the 2^{w-1} NEGATOR gates of the previous section all equal a classical gate: either an IDENTITY gate or a NOT gate.

Next, we proceed with proving that D is also a permutation matrix. The matrices V_{11} and V_{12} are complex permutation matrices. The matrix V_{11} contains $n/2$ non-zero entries. Among them, $n/2 - \mu_{11}$ can be chosen arbitrarily, μ_{11} being the weight of U_{11}. We denote these arbitrary numbers by x_j, analogous to x in the above example (5.1). Analogously, we denote by y_k the $n/2 - \mu_{12}$ arbitrary entries of V_{12}. Because U is a permutation matrix, the weight sum $\mu_{11} + \mu_{12}$ necessarily equals $n/2$. The matrix $-iV_{11}^\dagger V_{12}$ in (4.11) also is a complex permutation matrix and thus has $n/2$ non-zero entries. This number matches the total number of degrees of freedom $(n/2 - \mu_{11}) + (n/2 - \mu_{12}) = n/2$. Because U is a permutation matrix, V_{11} and V_{12} can be chosen such that the non-zero entries of the product $-iV_{11}^\dagger V_{12}$ depend only on an x_j or on an y_k but not on both. More particularly these entries are either of the form $-i/x_j$ or of the form $-iy_k$. By choosing all x_j equal to $-i$ and all y_k equal to i, the matrix $-iV_{11}^\dagger V_{12}$, and thus D, is a permutation matrix.

Because U, $\frac{1}{2} \left(\begin{smallmatrix} I+C & I-C \\ I-C & I+C \end{smallmatrix} \right)$, and $\left(\begin{smallmatrix} I & \\ & D \end{smallmatrix} \right)$ are permutation matrices, $\left(\begin{smallmatrix} A & \\ & B \end{smallmatrix} \right)$ also is an $n \times n$ permutation matrix. Ergo, given an $n \times n$ permutation matrix U, the Führ–Rzeszotnik procedure allows us to construct a 3-sandwich of permutation matrices. Therefore, we recover here the Birkhoff decomposition method for permutation matrices and thus, for $n = 2^w$, the primal synthesis method for classical reversible logic circuits in Chapter 2.

In an analogous way, application of (4.12) instead of (4.11) also yields a permutation-matrix decomposition into three permutation matrices. And there is more: we obtain exactly the same decomposition. Indeed, if we start from a permutation matrix U, then $(P_{11} + iP_{12})^2$ is the same diagonal matrix as $(P_{11} - iP_{12})^2$ and substituting $x_j = i$ and $y_k = -i$ into $iV_{11}^\dagger V_{12}$ gives the same matrix as substituting $x_j = -i$ and $y_k = i$ into $-iV_{11}^\dagger V_{12}$. We conclude that the two primal decompositions of a quantum circuit have one and the same primal decomposition of a classical circuit as a special case.

As an extra example, we consider the 8×8 permutation matrix

$$U = \begin{pmatrix} 0 & 0 & 0 & 1 & 0 & 0 & 0 & 0 \\ 0 & 0 & 0 & 0 & 0 & 0 & 0 & 1 \\ 0 & 0 & 0 & 0 & 0 & 0 & 1 & 0 \\ 0 & 0 & 0 & 0 & 0 & 1 & 0 & 0 \\ 1 & 0 & 0 & 0 & 0 & 0 & 0 & 0 \\ 0 & 0 & 0 & 0 & 1 & 0 & 0 & 0 \\ 0 & 0 & 1 & 0 & 0 & 0 & 0 & 0 \\ 0 & 1 & 0 & 0 & 0 & 0 & 0 & 0 \end{pmatrix}. \tag{5.2}$$

We have the following four polar decompositions:

$$U_{11} = \begin{pmatrix} 0 & 0 & 0 & 1 \\ 0 & 0 & 0 & 0 \\ 0 & 0 & 0 & 0 \\ 0 & 0 & 0 & 0 \end{pmatrix} = \begin{pmatrix} 1 & 0 & 0 & 0 \\ 0 & 0 & 0 & 0 \\ 0 & 0 & 0 & 0 \\ 0 & 0 & 0 & 0 \end{pmatrix} \begin{pmatrix} 0 & 0 & 0 & 1 \\ x_1 & 0 & 0 & 0 \\ 0 & x_2 & 0 & 0 \\ 0 & 0 & x_3 & 0 \end{pmatrix},$$

$$U_{12} = \begin{pmatrix} 0 & 0 & 0 & 0 \\ 0 & 0 & 0 & 1 \\ 0 & 0 & 1 & 0 \\ 0 & 1 & 0 & 0 \end{pmatrix} = \begin{pmatrix} 0 & 0 & 0 & 0 \\ 0 & 1 & 0 & 0 \\ 0 & 0 & 1 & 0 \\ 0 & 0 & 0 & 1 \end{pmatrix} \begin{pmatrix} y_1 & 0 & 0 & 0 \\ 0 & 0 & 0 & 1 \\ 0 & 0 & 1 & 0 \\ 0 & 1 & 0 & 0 \end{pmatrix},$$

$$U_{21} = \begin{pmatrix} 1 & 0 & 0 & 0 \\ 0 & 0 & 0 & 0 \\ 0 & 0 & 1 & 0 \\ 0 & 1 & 0 & 0 \end{pmatrix} = \begin{pmatrix} 1 & 0 & 0 & 0 \\ 0 & 0 & 0 & 0 \\ 0 & 0 & 1 & 0 \\ 0 & 0 & 0 & 1 \end{pmatrix} \begin{pmatrix} 1 & 0 & 0 & 0 \\ 0 & 0 & 0 & z_1 \\ 0 & 0 & 1 & 0 \\ 0 & 1 & 0 & 0 \end{pmatrix}, \text{ and}$$

$$U_{22} = \begin{pmatrix} 0 & 0 & 0 & 0 \\ 1 & 0 & 0 & 0 \\ 0 & 0 & 0 & 0 \\ 0 & 0 & 0 & 0 \end{pmatrix} = \begin{pmatrix} 0 & 0 & 0 & 0 \\ 0 & 1 & 0 & 0 \\ 0 & 0 & 0 & 0 \\ 0 & 0 & 0 & 0 \end{pmatrix} \begin{pmatrix} 0 & w_1 & 0 & 0 \\ 1 & 0 & 0 & 0 \\ 0 & 0 & w_2 & 0 \\ 0 & 0 & 0 & w_3 \end{pmatrix}.$$

We note that, in each of the four polar decompositions $U_{jk} = P_{jk} V_{jk}$, a zero eigenvalue of the positive semidefinite P_{jk} is accompanied by a free parameter in the unitary V_{jk}. Choosing $x_1 = x_2 = x_3 = -i$, $y_1 = i$, $z_1 = i$, and $w_1 = i$, and applying (4.11), leads to the primal decomposition

$$
\begin{pmatrix}
0 & 0 & 0 & 1 & & & & \\
1 & 0 & 0 & 0 & & & & \\
0 & 1 & 0 & 0 & & & & \\
0 & 0 & 1 & 0 & & & & \\
& & & & 1 & 0 & 0 & 0 \\
& & & & 0 & 0 & 0 & 1 \\
& & & & 0 & 0 & 1 & 0 \\
& & & & 0 & 1 & 0 & 0
\end{pmatrix}
\begin{pmatrix}
0 & & & 1 & & & & \\
& 0 & & & 1 & & & \\
& & 0 & & & 1 & & \\
& & & 1 & & & & 0 \\
1 & & & & 0 & & & \\
1 & & & & & 0 & & \\
1 & & & & & & 0 & \\
& & & 0 & & & & 1
\end{pmatrix}
\begin{pmatrix}
1 & 0 & 0 & 0 & & & & \\
0 & 1 & 0 & 0 & & & & \\
0 & 0 & 1 & 0 & & & & \\
0 & 0 & 0 & 1 & & & & \\
& & & & 0 & 0 & 0 & 1 \\
& & & & 0 & 0 & 1 & 0 \\
& & & & 0 & 1 & 0 & 0 \\
& & & & 1 & 0 & 0 & 0
\end{pmatrix}.
$$

5.4   GROUP HIERARCHY

As mentioned in Section 4.3, there are $\infty^{n^2/4}$ matrices in $\mathrm{bXU}(n)$. We investigate which of these bXU matrices are permutation matrices. For a matrix of form (4.6) to be a permutation matrix, the unitary matrix C is not allowed to have any non-zero elements outside its diagonal. Indeed, if $C_{jk} \neq 0$ with $j \neq k$, then the j th row of (4.6) would have four non-zero elements, a property not allowed for permutation matrices. The diagonal elements C_{jj} have to equal either 1 or -1. Otherwise, the j th row of (4.6) would have two non-zero elements. Hence, the unitary matrix C has to be an $n/2 \times n/2$ diagonal matrix with exclusively ± 1 entries. There exist only $2^{n/2}$ such matrices.

We now investigate which of the bZU matrices are permutation matrices. For a matrix of form (4.7) to be a permutation matrix, the unitary matrix D has to be an $n/2 \times n/2$ permutation matrix. There exist $(n/2)!$ such matrices.

We conclude: whereas

- all $n!$ elements of $\mathrm{P}(n)$ are inside the group $\mathrm{XU}(n)$ and

- only one element of $\mathrm{P}(n)$, i.e., the $n \times n$ unit matrix, is inside $\mathrm{ZU}(n)$,

there are

- only $2^{n/2}$ elements of $\mathrm{P}(n)$ inside $\mathrm{bXU}(n)$ and

- $(n/2)!$ elements of $\mathrm{P}(n)$ inside $\mathrm{bZU}(n)$,

- the remaining $n! - 2^{n/2} - (n/2)! + 1$ elements being neither in $\mathrm{bXU}(n)$ nor in $\mathrm{bZU}(n)$.

For example, the following P(4) matrices

$$
\begin{pmatrix}
 & 1 & & \\
 & & 1 & \\
 & & & 1 \\
1 & & &
\end{pmatrix},
\begin{pmatrix}
1 & & & \\
 & 1 & & \\
 & & & 1 \\
 & & 1 &
\end{pmatrix}, \text{ and }
\begin{pmatrix}
 & 1 & & \\
1 & & & \\
 & & & 1 \\
 & & 1 &
\end{pmatrix},
$$

respectively, belong

- to bXU(4) with $C = \begin{pmatrix} 1 & 0 \\ 0 & -1 \end{pmatrix}$,

- to bZU(4) with $D = \begin{pmatrix} 0 & 1 \\ 1 & 0 \end{pmatrix}$, and

- to neither of them.

All three belong to XU(4); see Figures 5.1 and 5.2.

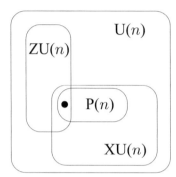

Figure 5.1: Venn diagram of the Lie groups U(n), XU(n), and ZU(n) and the finite groups P(n) and $\mathbf{1}(n)$.

Note: the trivial group $\mathbf{1}(n)$ is represented by the bullet.

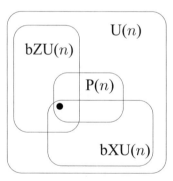

Figure 5.2: Venn diagram of the Lie groups U(n), bXU(n), and bZU(n) and the finite groups P(n) and $\mathbf{1}(n)$.

Note: the trivial group $\mathbf{1}(n)$ is represented by the bullet.

The intersection of the groups P(n) and bXU(n) is the group denoted P$_x$(n) in Section 2.2. It has order $2^{n/2}$ and is isomorphic to the direct product $\mathbf{S}_2^{n/2}$. A block XU(n) matrix can only be a permutation matrix provided C is a diagonal $n/2 \times n/2$ matrix with all $n/2$ diagonal entries equal to ± 1. The corresponding $n \times n$ matrices $\begin{pmatrix} I & \\ & C \end{pmatrix}$ can be interpreted as being built from $n/2$ blocks of size 2×2 and of type $\begin{pmatrix} 1 & \\ & \pm 1 \end{pmatrix}$. Therefore, if $n = 2^w$, such matrices represent multiply controlled Z gates acting on the first qubit. Thus the corresponding products $G \begin{pmatrix} I & \\ & C \end{pmatrix} G$ satisfy the equality

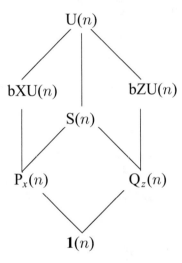

Analogously, the intersection of the groups $Q(n)$, defined in Section 3.8, and $bZU(n)$ is a group which we will denote $Q_z(n)$. It consists of all $n \times n$ diagonal matrices with exclusively entries 1 in the upper half diagonal and ± 1 in the lower half diagonal. It has order $2^{n/2}$ and is isomorphic to the direct product $\mathbf{S}_2^{n/2}$; see Figure 5.3.

Figure 5.3: Hierarchy of the Lie groups $U(n)$, $bXU(n)$, and $bZU(n)$ and the finite groups $S(n)$, $P_x(n)$, $Q_z(n)$, and $\mathbf{1}(n)$.

Although Figures 3.3 and 5.3 resemble each other, there is an important quantitative difference. Whereas in Figure 3.3 the groups $XU(n)$ and $ZU(n)$ have quite different dimensions (i.e., $(n-1)^2$ and $n-1$, respectively), in Figure 5.3 the groups $bXU(n)$ and $bZU(n)$ have the same dimension, i.e., $(n/2)^2$. Whereas in Figure 3.3 the groups $P(n)$ and $Q(n)$ have quite different orders (i.e., $n!$ and 2^{n-1}, respectively), in Figure 5.3 the groups $P_x(n)$ and $Q_z(n)$ have the same order, i.e., $2^{n/2}$. Thus, Figure 5.3 has a nice symmetry, absent in Figure 3.3. This symmetry is also expressed by the fact that $bXU(n)$ and $bZU(n)$ are conjugate and so are $P_x(n)$ and $Q_z(n)$:

$$bXU(n) = G\, bZU(n)\, G \quad \text{and} \quad bZU(n) = G\, bXU(n)\, G \; ;$$

$$P_x(n) = G \, Q_z(n) \, G \text{ and } Q_z(n) = G \, P_x(n) \, G \, ,$$

where G is the $n \times n$ matrix given by (4.4).

5.5   DUAL DECOMPOSITION

Each of the four blocks U_{jk} of a permutation matrix U simply consists of lines with:

- either all zeroes or

- zeroes and one 1.

The situation is more complicated for the conjugate matrix $u = GUG$, as the matrix u is not a permutation matrix. Therefore, its blocks u_{11}, u_{12}, u_{21}, and u_{22} are not $\{0, 1\}$ matrices. The block $u_{11} = \frac{1}{2}(U_{11} + U_{12} + U_{21} + U_{22})$ of u consists of lines with:

- either zeroes and one 1 or

- zeroes and twice $1/2$;

the blocks $u_{12} = \frac{1}{2}(U_{11} - U_{12} + U_{21} - U_{22})$ and $u_{21} = \frac{1}{2}(U_{11} + U_{12} - U_{21} - U_{22})$ consist of lines with:

- either zeroes or

- zeroes and twice $\pm 1/2$;

the block $u_{22} = \frac{1}{2}(U_{11} - U_{12} - U_{21} + U_{22})$ consists of lines with:

- either zeroes and one ± 1 or

- zeroes and twice $\pm 1/2$.

For the example (5.2), we have the following blocks u_{11}, u_{12}, u_{21}, and u_{22} and their polar decompositions:

$$u_{11} = \frac{1}{2} \begin{pmatrix} 1 & 0 & 0 & 1 \\ 1 & 0 & 0 & 1 \\ 0 & 0 & 2 & 0 \\ 0 & 2 & 0 & 0 \end{pmatrix} = \frac{1}{2} \begin{pmatrix} 1 & 1 & 0 & 0 \\ 1 & 1 & 0 & 0 \\ 0 & 0 & 2 & 0 \\ 0 & 0 & 0 & 2 \end{pmatrix} \frac{1}{2} \begin{pmatrix} 1+x_1 & 0 & 0 & 1-x_1 \\ 1-x_1 & 0 & 0 & 1+x_1 \\ 0 & 0 & 2 & 0 \\ 0 & 2 & 0 & 0 \end{pmatrix},$$

$$u_{12} = \frac{1}{2} \begin{pmatrix} 1 & 0 & 0 & 1 \\ -1 & 0 & 0 & -1 \\ 0 & 0 & 0 & 0 \\ 0 & 0 & 0 & 0 \end{pmatrix}$$

$$= \frac{1}{2} \begin{pmatrix} 1 & -1 & 0 & 0 \\ -1 & 1 & 0 & 0 \\ 0 & 0 & 0 & 0 \\ 0 & 0 & 0 & 0 \end{pmatrix} \frac{1}{2} \begin{pmatrix} 1+y_1 & 0 & 0 & 1-y_1 \\ -1+y_1 & 0 & 0 & -1-y_1 \\ 0 & 0 & 2y_2 & 0 \\ 0 & 2y_3 & 0 & 0 \end{pmatrix},$$

$$u_{21} = \frac{1}{2} \begin{pmatrix} -1 & 0 & 0 & 1 \\ -1 & 0 & 0 & 1 \\ 0 & 0 & 0 & 0 \\ 0 & 0 & 0 & 0 \end{pmatrix} = \frac{1}{2} \begin{pmatrix} 1 & 1 & 0 & 0 \\ 1 & 1 & 0 & 0 \\ 0 & 0 & 0 & 0 \\ 0 & 0 & 0 & 0 \end{pmatrix} \frac{1}{2} \begin{pmatrix} -1-z_1 & 0 & 0 & 1-z_1 \\ -1+z_1 & 0 & 0 & 1+z_1 \\ 0 & 0 & 2z_2 & 0 \\ 0 & 2z_3 & 0 & 0 \end{pmatrix},$$

and

$$
\begin{aligned}
u_{22} &= \frac{1}{2} \begin{pmatrix} -1 & 0 & 0 & 1 \\ 1 & 0 & 0 & -1 \\ 0 & 0 & -2 & 0 \\ 0 & -2 & 0 & 0 \end{pmatrix} \\
&= \frac{1}{2} \begin{pmatrix} 1 & -1 & 0 & 0 \\ -1 & 1 & 0 & 0 \\ 0 & 0 & 2 & 0 \\ 0 & 0 & 0 & 2 \end{pmatrix} \frac{1}{2} \begin{pmatrix} -1-w_1 & 0 & 0 & 1-w_1 \\ 1-w_1 & 0 & 0 & -1-w_1 \\ 0 & 0 & -2 & 0 \\ 0 & -2 & 0 & 0 \end{pmatrix}.
\end{aligned}
$$

Again, zero eigenvalues of the positive semidefinite factor are accompanied by degrees of freedom in the unitary factor. If $x_1 = \pm i$, $y_1 = -x_1$, $y_2 = \pm i$, $y_3 = \pm i$, $z_1 = -x_1$, $z_2 = \pm i$, and $z_3 = \pm i$, then we obtain a dual decomposition with exclusively permutation matrices. Choosing, in particular, $x_1 = z_2 = z_3 = i$ and $y_2 = y_3 = -i$, we find

$$
\begin{pmatrix} 1 & & & & 0 & & & \\ & 0 & & & & 1 & & \\ & & 1 & & & & 0 & \\ & & & 1 & & & & 0 \\ 0 & & & & 1 & & & \\ & 1 & & & & 0 & & \\ & & 0 & & & & 1 & \\ & & & 0 & & & & 1 \end{pmatrix}
\begin{pmatrix} 0 & 0 & 0 & 1 & & & & \\ 1 & 0 & 0 & 0 & & & & \\ 0 & 0 & 1 & 0 & & & & \\ 0 & 1 & 0 & 0 & & & & \\ & & & & 1 & 0 & 0 & 0 \\ & & & & 0 & 0 & 0 & 1 \\ & & & & 0 & 0 & 1 & 0 \\ & & & & 0 & 1 & 0 & 0 \end{pmatrix}
\begin{pmatrix} 0 & & & & 1 & & & \\ & 0 & & & & 1 & & \\ & & 0 & & & & 1 & \\ & & & 1 & & & & 0 \\ 1 & & & & 0 & & & \\ & 1 & & & & 0 & & \\ & & 1 & & & & 0 & \\ & & & 0 & & & & 1 \end{pmatrix}.
$$

In order to prove that a permutation matrix U always can be decomposed into three permutation matrices in the dual way, it would be sufficient to prove that, with the help of (4.13), we always can construct three matrices

$$\frac{1}{2} \begin{pmatrix} I + A' & I - A' \\ I - A' & I + A' \end{pmatrix}, \quad \begin{pmatrix} B' & \\ & C' \end{pmatrix}, \quad \text{and} \quad \frac{1}{2} \begin{pmatrix} I + D' & I - D' \\ I - D' & I + D' \end{pmatrix}$$

that are permutation matrices. Unfortunately, such a case is not possible, as is demonstrated by the example

$$U = \begin{pmatrix} 1 & 0 & 0 & 0 & 0 & 0 & 0 & 0 \\ 0 & 1 & 0 & 0 & 0 & 0 & 0 & 0 \\ 0 & 0 & 1 & 0 & 0 & 0 & 0 & 0 \\ 0 & 0 & 0 & 1 & 0 & 0 & 0 & 0 \\ 0 & 0 & 0 & 0 & 0 & 1 & 0 & 0 \\ 0 & 0 & 0 & 0 & 0 & 0 & 1 & 0 \\ 0 & 0 & 0 & 0 & 1 & 0 & 0 & 0 \\ 0 & 0 & 0 & 0 & 0 & 0 & 0 & 1 \end{pmatrix}. \tag{5.3}$$

Indeed, here we have $U_{12} = U_{21} = 0$ and

$$U_{11} = \begin{pmatrix} 1 & & & \\ & 1 & & \\ & & 1 & \\ & & & 1 \end{pmatrix} \quad \text{and} \quad U_{22} = \begin{pmatrix} 1 & & & \\ & 1 & & \\ & 1 & & \\ & & & 1 \end{pmatrix}.$$

The resulting matrices u_{11} and u_{12} have the following polar decompositions:

$$u_{11} = \frac{1}{2} \begin{pmatrix} 1 & 1 & 0 & 0 \\ 0 & 1 & 1 & 0 \\ 1 & 0 & 1 & 0 \\ 0 & 0 & 0 & 2 \end{pmatrix} = \frac{1}{6} \begin{pmatrix} 4 & 1 & 1 & 0 \\ 1 & 4 & 1 & 0 \\ 1 & 1 & 4 & 0 \\ 0 & 0 & 0 & 6 \end{pmatrix} \frac{1}{3} \begin{pmatrix} 2 & 2 & -1 & 0 \\ -1 & 2 & 2 & 0 \\ 2 & -1 & 2 & 0 \\ 0 & 0 & 0 & 3 \end{pmatrix}$$

and

$$u_{12} = \frac{1}{2} \begin{pmatrix} 1 & -1 & 0 & 0 \\ 0 & 1 & -1 & 0 \\ -1 & 0 & 1 & 0 \\ 0 & 0 & 0 & 0 \end{pmatrix}$$

$$= \frac{1}{2\sqrt{3}} \begin{pmatrix} 2 & -1 & -1 & 0 \\ -1 & 2 & -1 & 0 \\ -1 & -1 & 2 & 0 \\ 0 & 0 & 0 & 0 \end{pmatrix} \frac{1}{3} \begin{pmatrix} x+\sqrt{3} & x-\sqrt{3} & x & 0 \\ x & x+\sqrt{3} & x-\sqrt{3} & 0 \\ x-\sqrt{3} & x & x+\sqrt{3} & 0 \\ 0 & 0 & 0 & 3y \end{pmatrix}.$$

Applying (4.13) results in

$$B' = \frac{1}{2\sqrt{3}} \begin{pmatrix} \sqrt{3}+i & \sqrt{3}+i & -2i & 0 \\ -2i & \sqrt{3}+i & \sqrt{3}+i & 0 \\ \sqrt{3}+i & -2i & \sqrt{3}+i & 0 \\ 0 & 0 & 0 & 2\sqrt{3} \end{pmatrix},$$

a matrix which is unitary (and, in fact, an XU matrix), but unfortunately not a permutation matrix. The cause of this problem is the lack of degrees of freedom in W_{11}. The above matrix W_{11} is unique because $\det(u_{11}) = 1/4$ and thus is non-zero. A non-zero determinant of u_{11} appears whenever u_{11} contains exclusively $k \times k$ cycles G_k with odd k. Here, G_k is defined by $G_1 = (1)$ and, for $k > 1$, by

$$
G_k = \frac{1}{2}
\begin{pmatrix}
1 & 1 & & & & & \\
 & 1 & 1 & & & & \\
 & & 1 & 1 & & & \\
 & & & \ddots & \ddots & & \\
 & & & & & 1 & 1 \\
1 & & & & & & 1
\end{pmatrix}.
$$

Indeed, if k is even, then $\det(2G_k) = 0$ and thus the polar decomposition of G_k is not unique, but if k is odd, then $\det(2G_k) = 2$ and the decomposition is unique. Thus the conclusion of the present section is: a dual classical decomposition exists but cannot always be found by means of the quantum method of Section 4.4. It can be found by the classical method of Section 2.4.

Thus, whereas the primal decomposition of a permutation matrix according to the quantum algorithm of Section 4.2 always supplies a decomposition into permutation matrices, the dual decomposition of a permutation matrix according to the quantum algorithm of Section 4.4 does not always supply a decomposition into permutation matrices. If the $n \times n$ matrix U is a permutation matrix, then we can always get rid of the two HADAMARD gates in (4.5), but not always the four HADAMARD gates in (4.10).

CHAPTER 6

Conclusion

We summarize the previous chapters:

- In Chapter 2, we presented three synthesis methods for classical reversible circuits, based on three decompositions of an arbitrary permutation matrix, with increasing efficiency:

  - a primal matrix decomposition,
  - a dual matrix decomposition, and
  - a refined matrix decomposition.

  They lead to three circuits, each composed of exclusively controlled NOT gates. The third method needs only $2w - 1$ such gates (w being the width of the circuit, i.e., the number of processed bits) and therefore is almost optimally efficient.

- In Chapter 3, we replaced controlled NOT gates by controlled NEGATOR gates and controlled PHASOR gates. This enables us to step from classical computing to quantum computing.

- In Chapter 4, we presented two synthesis methods for quantum circuits, based on two decompositions of an arbitrary unitary matrix:

  - a primal matrix decomposition and
  - a dual matrix decomposition.

  Both methods are optimally efficient, as they need (besides HADAMARD gates) only $(2^w)^2$ gates, i.e., as many as there are degrees of freedom in the unitary matrix.

- In Chapter 5, we demonstrated how

  - the primal unitary decomposition of Chapter 4 contains the primal permutation decomposition of Chapter 2 as a special case, but
  - the dual unitary decomposition of Chapter 4 does not contain the dual permutation decomposition of Chapter 2.

All chapters successfully take advantage of group theory, both of finite groups (classical computing) and Lie groups (quantum computing). The relevant groups follow a definite hierarchy: in Chapter 3, we have constructed the group graph of Figure 3.3; in Chapter 5, we have constructed the group graph of Figure 5.3. Merging both figures gives Figure 6.1. We note that the symmetry in Figure 6.1 is broken: whereas we have an edge between the vertices bXU(n) and XU(n), we lack an edge between the vertices bZU(n) and ZU(n). Indeed, XU(n) is a supergroup of bXU(n), but ZU(n) is not a supergroup of bZU(n).

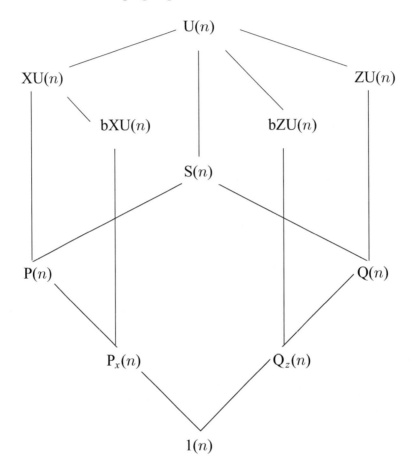

Figure 6.1: Hierarchy of the Lie groups U(n), XU(n), ZU(n), bXU(n), and bZU(n) and the finite groups S(n), P(n), Q(n), P_x(n), Q_z(n), and $\mathbf{1}$(n).

In order to better illustrate the relation between the quantum case (i.e., the unitary-matrix case) and the classical case (i.e., the permutation-matrix case), we recall here (for even n) two subgroups of P(n). For this purpose, we consider the $n \times n$ permutation matrix P as composed of four $n/2 \times n/2$ blocks:

$$P = \left(\begin{array}{cc} P_{11} & P_{12} \\ P_{21} & P_{22} \end{array} \right) .$$

The two subgroups are defined as follows:

- the group $P_x(n)$ consists of the permutation matrices with all four blocks P_{11}, P_{12}, P_{21}, and P_{22} diagonal and

- the group $P_z(n)$ consists of the permutation matrices with block P_{11} equal to the $n/2 \times n/2$ unit matrix I.

The group $P_x(n)$ is already mentioned in Sections 2.2 and 5.4. It equals $P(n) \cap bXU(n)$. The group $P_z(n)$ is mentioned in Section 2.2 and equals $P(n) \cap bZU(n)$. From Chapter 2 it is clear that the closure of $P_x(n)$ and $P_z(n)$ equals $P(n)$. Figure 6.2 (fusion of Figures 2.3 and 4.1) shows the group hierarchy. The graph is symmetric. However, this symmetry is broken when we look at the group orders: Figure 6.3. Whereas $bXU(n)$ and $bZU(n)$ have the same dimension, $P_x(n)$ has lower order than $P_z(n)$. The reason is as follows. If C is an arbitrary $n/2 \times n/2$ unitary matrix, then automatically $\frac{1}{2} \begin{pmatrix} I+C & I-C \\ I-C & I+C \end{pmatrix}$ is also an $n \times n$ unitary matrix. In contrast, if C is an arbitrary $n/2 \times n/2$ permutation matrix, then $\frac{1}{2} \begin{pmatrix} I+C & I-C \\ I-C & I+C \end{pmatrix}$ is not an $n \times n$ permutation matrix (except if $C = I$).

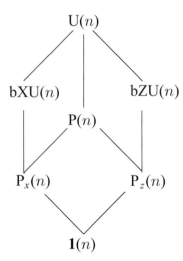

Figure 6.2: Hierarchy of the Lie groups $U(n)$, $bXU(n)$, and $bZU(n)$ and the finite groups $P(n)$, $P_x(n)$, $P_z(n)$, and $\mathbf{1}(n)$.

The symmetry between $bXU(n)$ and $bZU(n)$ is the reason why the quantum primal and dual synthesis methods have the same efficiency; the asymmetry between $P_x(n)$ and $P_z(n)$ is the reason why the classical dual synthesis method is more efficient than the classical primal synthesis method and also the reason why a "refined" method performs even better than the dual method.

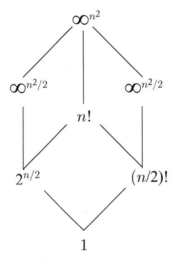

Figure 6.3: Orders of the Lie groups U(n), bXU(n), and bZU(n) and the finite groups P(n), P$_x$(n), P$_z$(n), and $\mathbf{1}$(n).

APPENDIX A

Polar Decomposition

If both a and b are arbitrary real numbers, then the complex number $c = a + ib$ can be written as a product

$$c = pw ,$$

where p is a real number, either zero or positive, and w is a complex number with unit modulus and thus of the form $e^{i\theta}$ with θ real. If $c \neq 0$, then both p and w are unique. If, however, $c = 0$, then $p = 0$ and w may be any number on the unit circle of the complex plane and θ thus may be any real number.

The generalization of this factorization from a number c to a square matrix C (with complex entries) is called the polar decomposition of C: any $n \times n$ matrix C can be decomposed as follows:

$$C = PW ,$$

where P is an $n \times n$ positive semidefinite matrix and W is an $n \times n$ unitary matrix. The matrix P is unique. If C is invertible, then also W is unique. If, however, C is singular, then W has degrees of freedom.

We give here an example of the polar decomposition of an invertible 2×2 matrix:

$$\begin{pmatrix} 1 & 0 \\ i & -i \end{pmatrix} = \frac{1}{\sqrt{5}} \begin{pmatrix} 2 & -i \\ i & 3 \end{pmatrix} \frac{1}{\sqrt{5}} \begin{pmatrix} 2 & 1 \\ i & -2i \end{pmatrix} .$$

For finding the unitary factor W of the polar decomposition PW of a given square matrix C, we proceed in the following iterative way, called Hero's method [64]:

$$\begin{aligned} C_0 &= C \\ C_{j+1} &= \frac{1}{2} [\, C_j + (C_j^{-1})^\dagger \,] . \end{aligned}$$

Once an approximation C_k for W is found, CC_k^{-1} is the corresponding approximation for P. For example, after three iterations, we obtain

$$\begin{pmatrix} 1 & 0 \\ i & -i \end{pmatrix} \approx \begin{pmatrix} 0.8944099 & -0.4472049\,i \\ 0.4472049\,i & 1.3416149 \end{pmatrix} \begin{pmatrix} 0.8944444 & 0.4472222 \\ 0.4472222\,i & -0.8944444\,i \end{pmatrix} .$$

As the mathematician Hero (a.k.a. Heron) lived in Alexandria, Egypt, around the years 100 or 200, he, of course, had no idea what a unitary matrix was, let alone how to compute the polar decomposition of a square matrix. Nevertheless, the iterative procedure is named after him, because he invented the following algorithm for computing approximations for the square root x of 2:

$$
\begin{aligned}
x_0 &= 1 \\
x_{j+1} &= \frac{1}{2} \left[\, x_j + 2/x_j \,\right].
\end{aligned}
$$

Already after three iterations, this procedure yields a very good rational approximation $x_3 = 577/408 \approx 1.4142156$ of the irrational number $\sqrt{2} \approx 1.4142135$.

Unfortunately, Hero's method only works if the given matrix C is invertible. If, on the contrary, C is singular, then we have to return to some other method to determine the factors P and W. This may happen by means of the singular-value decomposition of C. One also may replace the singular matrix C by a regular matrix C', "almost equal" to C.

Bibliography

[1] T. Bartee, *Digital Computer Fundamentals*, McGraw–Hill Book Company, New York, pp. 66–151, 1966. 1

[2] G. Casanova, *L'algèbre de Boole*, Presses Universitaires de France, Paris, 1967. 1

[3] T. Sasao and M. Fujita, *Representation of Discrete Functions*, Kluwer Academic Publishers, Boston, 1996. DOI: 10.1007/978-1-4613-1385-4. 1, 4, 6

[4] A. Gaidukov, Algorithm to derive minimum ESOP for 6-variable function, *Proc. of the 5th International Workshop on Boolean Problems*, Freiberg, pp. 141–148, September 2002. 6, 7

[5] G. Pogosyan, I. Rosenberg, and S. Takada, Building minimum ESOPs through redundancy elimination, *Proc. of the 6th International Workshop on Boolean Problems*, Freiberg, pp. 201–206, September 2004. 6

[6] E. Dubrova, D. Miller, and J. Muzio, Upper bound on number of products in AND-OR-XOR expansions of logic functions, *Electronics Letters*, vol. 31, pp. 541–542, 1995. DOI: 10.1049/el:19950377. 7

[7] The GAP group, GAP—a tutorial, https://www.gap-system.org/Manuals/doc/tut/manual.pdf, 2016. 10

[8] E. Fredkin and T. Toffoli, Conservative logic, *International Journal of Physics*, vol. 21, pp. 219–253, 1982. DOI: 10.1007/bf01857727. 10

[9] A. De Vos, *Reversible Computing*, Wiley, VCH, Weinheim, 2010. DOI: 10.1002/9783527633999. 14, 21, 40, 44, 65

[10] R. Wille and R. Drechsler, *Towards a Design Flow for Reversible Logic*, Springer, Dordrecht, 2010. DOI: 10.1007/978-90-481-9579-4. 10, 56

[11] M. Soeken, R. Wille, O. Keszöcze, M. Miller, and R. Drechsler, Embedding of large Boolean functions for reversible logic, *ACM Journal on Emerging Technologies in Computing Systems*, vol. 12, p. 41, 2015. DOI: 10.1145/2786982. 10

[12] A. De Vos and Y. Van Rentergem, Young subgroups for reversible computers, *Advances in Mathematics of Communications*, vol. 2, pp. 183–200, 2008. DOI: 10.3934/amc.2008.2.183. 14, 34, 35, 37, 40

[13] A. Yong, What is ... a Young tableau?, *Notices of the AMS*, vol. 54, pp. 240–241, 2007. 14

[14] G. Birkhoff, Tres observaciones sobre el algebra lineal, *Universidad Nacional de Tucumán: Revista Matemáticas y Física Teórica*, vol. 5, pp. 147–151, 1946. 15

[15] R. Bhatia, *Matrix Analysis*, Springer, New York, 1997. DOI: 10.1007/978-1-4612-0653-8. 15

[16] D. de Werra, Path coloring in bipartite graphs, *European Journal of Operational Research*, vol. 164, pp. 575–584, 2005. DOI: 10.1016/j.ejor.2003.05.007. 15

[17] C. Peng, G. Bochman, and T. Hall, Quick Birkhoff-von Neumann decomposition algorithm for agile all-photonic network cores, *Proc. of the IEEE International Conference on Communications*, Istanbul, pp. 2593–2598, June 2006. DOI: 10.1109/icc.2006.255170. 15

[18] A. De Vos and Y. Van Rentergem, Reversible computing: From mathematical group theory to electronical circuit experiment, *Proc. of the Computing Frontiers Conference*, Ischia, pp. 35–44, May 2005. DOI: 10.1145/1062261.1062270. 17

[19] A. Kerber, *Representations of Permutation Groups I*, Springer Verlag, Berlin, pp. 17–23, 1970. DOI: 10.1007/bfb0067943. 18

[20] G. James and A. Kerber, The representation theory of the symmetric group, *Encyclopedia of Mathematics and its Applications*, vol. 16, pp. 15–33, 1981. DOI: 10.1017/cbo9781107340732.005.

[21] A. Jones, A combinatorial approach to the double cosets of the symmetric group with respect to Young subgroups, *European Journal of Combinatorics*, vol. 17, pp. 647–655, 1996. DOI: 10.1006/eujc.1996.0056. 18

[22] M. Nielsen and I. Chuang, *Quantum Computation and Quantum Information*, Cambridge University Press, Cambridge, 2000. DOI: 10.1017/cbo9780511976667. 21

[23] B. Eastin and S. Flammia, Q-Circuit tutorial, http://info.phys.unm.edu/Qcircuit/, 2008. 25

[24] A. De Vos and Y. Van Rentergem, Synthesis of reversible logic for nanoelectronic circuits, *International Journal of Circuit Theory and Applications*, vol. 35, pp. 325–341, 2007. DOI: 10.1002/cta.413. 31, 32, 34

[25] Y. Van Rentergem and A. De Vos, Synthesis and optimization of reversible circuits, *Proc. of the Reed–Muller Workshop*, Oslo, pp. 67–75, May 2007. 34

[26] A. De Vos and Y. Van Rentergem, Networks for reversible logic, *Proc. of the 8th International Workshop on Boolean Problems*, Freiberg, pp. 41–47, September 2008. 34

[27] L. Chen and L. Yu, Decomposition of bipartite and multipartite unitary gates, *Physical Review A*, vol. 91, 032308, 2015. DOI: 10.1103/physreva.91.032308. 35, 44

[28] Y. Van Rentergem, *Ontwerp van Reversibele Digitale Schakelingen*, Universiteit Gent, Gent, pp. 133–134, 2007. 37

[29] A. De Vos and Y. Van Rentergem, Multiple-valued reversible logic circuits, *Journal of Multiple-Valued Logic and Soft Computing*, vol. 15, pp. 489–505, 2009. 39

[30] C. Clos, A study of non-blocking switching networks, *Bell Systems Technical Journal*, vol. 32, pp. 406–424, 1953. DOI: 10.1002/j.1538-7305.1953.tb01433.x. 42

[31] F. Hwang, Control algorithms for rearrangeable Clos networks, *IEEE Transactions on Communications*, vol. 31, pp. 952–954, 1983. DOI: 10.1109/tcom.1983.1095923. 42

[32] J. Hui, *Switching and Traffic Theory for Integrated Broadband Networks*, Kluwer Academic Publishers, Boston, pp. 53–138, 1990. DOI: 10.1007/978-1-4615-3264-4.

[33] J. Chao, Z. Jing, and S. Liew, Matching algorithms for three-stage bufferless Clos network switches, *IEEE Communications Magazine*, vol. 41, pp. 46–54, 2003. DOI: 10.1109/mcom.2003.1235594. 42

[34] A. Jajszczyk, Nonblocking, repackable, and rearrangeable Clos networks: 50 years of the theory evolution, *IEEE Communications Magazine*, vol. 41, pp. 28–33, 2003. DOI: 10.1109/MCOM.2003.1235591. 42

[35] N. Abdessaied, M. Soeken, M. Thomsen, and R. Drechsler, Upper bounds for reversible circuits based on Young subgroups, *Information Processing Letters*, vol. 114, pp. 282–286, 2014. DOI: 10.1016/j.ipl.2014.01.003. 45

[36] M. Soeken, S. Frehse, R. Wille, and R. Drechsler, RevKit: An open source toolkit for the design of reversible circuits, *Proc. of the 3rd International Workshop on Reversible Computation*, Gent, pp. 64–76, July 2011. DOI: 10.1007/978-3-642-29517-1_6. 47

[37] M. Soeken, RevKit, msoeken.github.io/revkit.html, 2016. DOI: 10.1007/978-3-642-29517-1_6. 47

[38] D. Deutsch, Quantum computation, *Physics World*, vol. 5, no. 6, pp. 57–61, 1992. DOI: 10.1088/2058-7058/5/6/38. 49

[39] D. Deutsch, A. Ekert, and R. Lupacchini, Machines, logic and quantum physics, *The Bulletin of Symbolic Logic*, vol. 3, pp. 265–283, 2000. DOI: 10.2307/421056.

[40] A. Galindo and M. Martín-Delgado, Information and computation: Classical and quantum aspects, *Review of Modern Physics*, vol. 74, pp. 347–423, 2002. DOI: 10.1103/revmodphys.74.347.

102   BIBLIOGRAPHY

[41] D. Miller, Decision diagram techniques for reversible and quantum circuits, *Proc. of the 8th International Workshop on Boolean Problems*, Freiberg, pp. 1–15, September 2008.

[42] S. Vandenbrande, R. Van Laer, and A. De Vos, The computational power of the square root of NOT, *Proc. of the 10th International Workshop on Boolean Problems*, Freiberg, pp. 257–262, September 2012. 49

[43] A. De Vos, J. De Beule, and L. Storme, Computing with the square root of NOT, *Serdica Journal of Computing*, vol. 3, pp. 359–370, 2009. 51, 52

[44] O. Elgerd, *Control Systems Theory*, McGraw–Hill Book Company, New York, pp. 384–411, 1967. 52

[45] A. De Vos, R. Van Laer, and S. Vandenbrande, The group of dyadic unitary matrices, *Open Systems and Information Dynamics*, vol. 19, 1250003, 2012. DOI: 10.1142/s1230161212500035. 52

[46] W. Castryck, J. Demeyer, A. De Vos, O. Keszöcze, and M. Soeken, Translating between the roots of the identity in quantum computers, *Proc. of the 48th International Symposium on Multiple-Valued Logic*, Linz, pp. 254–259, May 2018. 53

[47] A. De Vos and S. De Baerdemacker, Matrix calculus for classical and quantum circuits, *ACM Journal on Emerging Technologies in Computing Systems*, vol. 11, p. 9, 2014. DOI: 10.1145/2669370. 55, 56

[48] Z. Sasanian and D. Miller, Transforming MCT circuits to NCVW circuits, *Proc. of the 3rd International Workshop on Reversible Computation*, Gent, pp. 163–174, July 2011. DOI: 10.1007/978-3-642-29517-1_7. 56

[49] P. Selinger, Efficient Clifford $+T$ approximations of single-qubit operators, *Quantum Information and Computation*, vol. 15, pp. 159–180, 2015.

[50] M. Amy, D. Maslov, and M. Mosca, Polynomial-time T-depth optimization of Clifford$+T$ circuits via matroid partitioning, *IEEE Transactions on Computer-Aided Design of Integrated Circuits and Systems*, vol. 33, 1486, 2013. DOI: 10.1109/tcad.2014.2341953. 56

[51] A. De Vos and S. De Baerdemacker, Scaling a unitary matrix, *Open Systems and Information Dynamics*, vol. 21, 1450013, 2014. DOI: 10.1142/s1230161214500139. 56, 63, 64

[52] A. De Vos and S. De Baerdemacker, On two subgroups of U(n), useful for quantum computing, *Journal of Physics: Conference Series: Proceedings of the 30th International Colloquium on Group-theoretical Methods in Physics, Gent, (July 2014)*, vol. 597, 012030, 2015. DOI: 10.1088/1742-6596/597/1/012030. 56

[53] D. Bouwmeester and A. Zeilinger, The physics of quantum information: Basic concepts, In: D. Bouwmeester, A. Ekert, and A. Zeilinger, *The Physics of Quantum Information*, Springer Verlag, Berlin, pp. 1–14, 2000. DOI: 10.1007/978-3-662-04209-0. 57

[54] A. De Vos and S. De Baerdemacker, The NEGATOR as a basic building block for quantum circuits, *Open Systems and Information Dynamics*, vol. 20, 1350004, 2013. DOI: 10.1142/s1230161213500042. 59, 62

[55] A. Hurwitz, Ueber die Erzeugung der Invarianten durch Integration, *Nachrichten von der Königliche Gesellschaft der Wissenschaften zu Göttingen, Mathematisch-physikalische Klasse*, vol. 1897, pp. 71–90, 1897. DOI: 10.1007/978-3-0348-4160-3_38. 59

[56] M. Poźniak, K. Życzkowski, and M. Kuś, Composed ensembles of random unitary matrices, *Journal of Physics A: Mathematical and General*, vol. 31, pp. 1059–1071, 1998. DOI: 10.1088/0305-4470/31/3/016. 59, 76

[57] R. Jozsa, Quantum algorithms, In: D. Bouwmeester, A. Ekert, and A. Zeilinger, *The Physics of Quantum Information*, Springer Verlag, Berlin, pp. 104–126, 2000. 62

[58] M. Idel and M. Wolf, Sinkhorn normal form for unitary matrices, *Linear Algebra and its Applications*, vol. 471, pp. 76–84, 2015. DOI: 10.1016/j.laa.2014.12.031. 63, 69

[59] R. Sinkhorn, A relationship between arbitrary positive matrices and doubly stochastic matrices, *Annals of Mathematical Statistics*, vol. 35, pp. 876–879, 1964. DOI: 10.1214/aoms/1177703591. 64

[60] T. Beth and M. Rötteler, Quantum algorithms: Applicable algebra and quantum physics, In: G. Alber, T. Beth, M. Horodecki, P. Horodecki, R. Horodecki, M. Rötteler, H. Weinfurter, R. Werner, and A. Zeilinger, *Quantum Information*, Springer Verlag, Berlin, pp. 96–150, 2001. DOI: 10.1007/3-540-44678-8. 65

[61] A. De Vos and S. De Baerdemacker, The synthesis of a quantum circuit, In: B. Steinbach, *Problems and New Solutions in the Boolean Domain*, Cambridge Scholars Publishing, pp. 357–368, 2016. 69

[62] H. Führ and Z. Rzeszotnik, On biunimodular vectors for unitary matrices, *Linear Algebra and its Applications*, vol. 484, pp. 86–129, 2015. DOI: 10.1016/j.laa.2015.06.019. 70, 74

[63] A. De Vos and S. De Baerdemacker, Block-ZXZ synthesis of an arbitrary quantum circuit, *Physical Review A*, vol. 94, 052317, 2016. DOI: 10.1103/physreva.94.052317. 73, 74

[64] N. Higham, Computing the polar decomposition with applications, *SIAM Journal on Scientific and Statistical Computing*, vol. 7, pp. 1160–1174, 1986. DOI: 10.1137/0907079. 75, 97

[65] K. Życzkowski and M. Kuś, Random unitary matrices, *Journal of Physics A: Mathematical and General*, vol. 27, pp. 4235–4245, 1994. DOI: 10.1088/0305-4470/27/12/028. 76

Authors' Biographies

ALEXIS DE VOS

Alexis De Vos is an electrical engineer, physicist, and doctor in applied sciences and graduated from the Universiteit Gent (Belgium). He is currently a part-time professor in the Department of Electronics of the Universiteit Gent. His research is concerned with material science (polymers, semiconductors, metals, liquid crystals), microelectronics (thin films, chips, neural networks, reversible circuits), and energy sciences (thermodynamics, solar energy, endoreversible engines). He is author of the books *Thermodynamics of Solar Energy Conversion* (Wiley-VCH, 2008) and *Reversible Computing* (Wiley-VCH, 2010). He designed and produced several prototype integrated circuits for reversible computers such as adders, multipliers, and linear transformers. He currently investigates quantum computing.

STIJN DE BAERDEMACKER

Stijn De Baerdemacker is a physicist and doctor in sciences and graduated from the Universiteit Gent (Belgium). He has been a visiting scientist at the University of Toronto (ON, Canada), University of Notre Dame (IN, USA) and Universiteit Amsterdam (The Netherlands). He is currently a post-doctoral researcher in the Department of Physics and Astronomy of the Universiteit Gent. His research is concerned with the development of accurate quantum many-body methods in quantum physics, quantum chemistry and quantum computing. For this, he uses and develops techniques from Lie algebra theory and notions from (quantum) integrability. In his free time, he is a painter and explores the boundaries between science and art.

YVAN VAN RENTERGEM

Yvan Van Rentergem is an electrical engineer and doctor in applied sciences and graduated from the Universiteit Gent (Belgium). He obtained his Ph.D. in 2008 in the subject of reversible computing. During his research, he developed algorithms for the synthesis of reversible circuits. These methods were applied for real-life prototype chips. His research led to ten articles presented at international conferences or published in international journals. After earning his Ph.D., he went to work at ArcelorMittal Gent as operations research specialist, developing models to optimize the logistical flow of the steel shop. These models are successfully applied at ArcelorMittal Gent and several other sites of ArcelorMittal. He currently is slab yard support manager at ArcelorMittal Gent.

Index

Printed in the United States
by Baker & Taylor Publisher Services